IF NOT GOD, THEN WHAT?

Neuroscience, Aesthetics, and the
Origins of the Transcendent

JOSHUA FOST

A *Vox Ratio* Publication

For Quincy

First published 2007 by Clearhead Studios, Inc. For further information on Clearhead Studios, write to info@clearheadstudios.com or visit our website: http://www.clearheadstudios.com/

ISBN 978-0-6151-6106-8

Designed by Joshua Fost

CONTENTS

PREFACE

ew domains of human activity have achieved the longevity and influence of religion. We probably invented (some might say *discovered*) some form of it around the time we started making tools. For hundreds of millions of people, religion addresses all of the most important questions of their lives. Where do we come from? Where are we going? How should we live? Perhaps the only other force that has had as big an influence is science. Its technological fruits also affect almost every aspect of our lives, from how we're born, to how we live, and how we die. And if we think of science as an effort to understand nature, then it too goes back, in some form, to our origins as a species.

Isn't it odd, then, that these two forces have so little to do with each other in the modern world? What exactly is the difference between them? Do they both not seek to make sense of experience? The party line answer, of course, is that religion deals with the intangible aspects of life (the "why") while science handles the material aspects (the "how"). While religion answers questions of the spirit and of ethics, science answers questions of the body and its physical environment. The two are supposed to complement each other. As the paleontologist Stephen Jay Gould said, they occupy "non-overlapping magisteria." Many people are perfectly comfortable with that model and see no problem entertaining both scientific and religious worldviews simultaneously. Darwin famously rejected that possibility, seeing deep conflict between the two "ways of knowing," and somewhat reluctantly, given his religious roots, started a turf war that persists even today, to the consternation and amazement of many. By providing a naturalistic theory of our bodies' origins, Darwin stepped on a religious toe. The Bible gave a competitive explanation for where we came from, and it certainly wasn't a swamp or a savannah.

While Darwin is rightly credited for his explanation of evolution's mechanism, he didn't start the fight between science and religion. Before him there was Galileo, and before him Copernicus. And then there are the methodological differences between faith and inquiry, elucidation of which goes all the way back, not particularly attributable to any one person. These deep tensions have led to a correspondingly long history of acrimony, but outright direct conflict seems to have a way of dissolving into the background until, typically, an upstart scientist begins to research something of particular interest to theists (although the table is occasionally turned when religionists try to tell scientists what they can and cannot study and teach – witness Galileo's house arrest, Giordano Bruno's execution, creationism in the biology classroom, and prohibitions against stem cell research). Few churches really care about whether *E. coli* can be tricked into making human insulin, so that work proceeds without much static, but if the research concerns the legal or ethical status of a human blastocyst, hackles rise.

This book is a bellwether that the border between science and religion has been crossed again, albeit in a more gradual, less discrete way. The science responsible for the border crossing is neuroscience, and churchgoers are taking (or should be taking) notice because the research concerns nothing less than the biological basis of our minds, even our souls. If a Darwinian origin for our opposable thumb offended the idea of divine creation, what can we expect from the neuroscientific suggestion that the traditional experience of God is a neuroelectrical storm? More generally, even when the research has nothing to do with religious experience *per se*, there is a latent sensitivity to reductionist accounts of psychology. Most believers see our minds as closer links to divinity than our bodies, so comparative anatomy gets a pass but comparative neuroethology does not. Many of the theists who accept evolution still have a hard time with the question of what happens to the immortal soul of someone whose personality is radically altered by a brain injury.

The prospect of answering these questions and alleviating some of the tension is not, however, cause for despair. Some of the same neuroscience that prompts these questions creates an opportunity to consolidate several vital aspects of human psychology into one. In this book, it's not just religious psychology that will be placed under a microscope, but "scientistic" psychology as well. In other words, we'll examine the possibility that the apparent connection between science and religion, as quests to understand experience, is not just apparent or superficial, but deeply fundamental. Many of the same neural processes that lead us to religious modes of thinking lead us also to scientific modes of thinking. By invoking a common mechanism for both sets of phenomena, neuroscience may still be responsible for firing the first shot across the border, but it will also deliver us from the fight by marrying, in subject matter if not methodology, these traditionally opposing forces.

This book is written mostly for two audiences: (1) religious moderates, and (2) scientists (popular science audiences included). Operating under the principle that a good compromise is one in which neither side is really very happy, there are sections that will make both groups squirm. Religionists may find the book's naturalist stance rather stark, while scientists may find the phenomena to be explained somewhat subjective. Non-scientist believers may find some of the biology or philosophy challenging, but faithless Ph.D.s may balk at the introspective mysticism. Such twists in the road are inevitable if religionists are to have their experience explained and scientists are to have their logical and empirical requirements met. On those two points, both groups should be satisfied. And irrespective of existing attitudes about these matters, any honest inquirer should walk away with a new perspective on familiar subjects.

Secondarily, and more generally, this book is written for those readers who care about the consequences of ideas. The United States is perhaps at once the most religious and the most scientifically advanced nation in the world. With the combined power of these institutions to shape our wills and our environment, we should take care to understand their intersection as clearly as possible. That intersection is worldview, so this book is an effort not just to shine a light on nature, but to explore the meanings and implications of that study for our everyday lives.

CHAPTER 1

The Search for Beauty

In 13th century Europe, the academic pursuit known as medieval Scholasticism began to combine theology with logical, mathematical, and natural investigations. Disputations on early medicine, philosophy, and geometry all merged in the influential and academic studies of the clergy – indeed William of Ockham, whom we will meet later, was himself a Franciscan scholastic. One way of thinking about the impulse behind this multidisciplinary trend is as a desire to study not just God Himself but also God's creation, and an attempt at reconciling the study of these hitherto distinct worlds. The results of Scholasticism were tremendously influential, as they led ultimately to the maturation of natural philosophy, from which modern science was born. The distinct methods, however, as well as the subject matter, that began to distinguish science from theology gradually separated the domains. By the time of the Enlightenment, clear delineations were evident, and it is a rare Franciscan monk who, in the 21st century, maintains an active research program in molecular biology. These days, for most people, most of the time, science and religion seem cleanly separated. The vestiges of the deeper connection sought by the

Scholastics lie in the modern consideration of contentious issues, some of which we will be discussing later, and also in low-membership tribal religions, where explanations of natural phenomena are inseparable from an understanding of the psychologies and caprices of pantheonic gods. To ensure good rains and healthy crops, for instance, believers in these religions celebrate feasts in honor of so-and-so in the spirit world.

This historical mixing of spirituality and empiricism shines light on deeply rooted psychological drives: The need for (1) an emotional connection to the perfection of the universe (interpreted as God or gods); and (2) an intellectual connection to the perfection of the universe (interpreted as Nature). In modern society, science has obviously concerned itself with the latter, but if we are ever going to achieve a total explanation of human experience, the emotional and spiritual side also demands a naturalist address. Although it is a contentious subject, excluding these aspects of human experience from the range of phenomena science explains would seem as arbitrary as excluding taste or smell or touch because they too have a subjective character.

The approach of this book will be to re-establish a sort of a union of these two quests, though in a way quite unimagined by the Scholastics. Ironically, that union is exactly what will extract us from the tangle and tension we suffer from today. Some scientists are quite hostile even to the discussion of emotional, or more pointedly *spiritual* connections to nature, perhaps because of an enduring legacy of behaviorism that internal subjective states are difficult to study (though see work by neuroscientist Joseph LeDoux for some major insights). A scientific challenge is one thing, but to say that emotional structures lie outside of science is dualism, *i.e.* that form of supernaturalism asserting that the human mind is a substance utterly unlike everything else in the rest of the universe, and as such is untenable. More practically, and disturbingly, scientific silence on this subject means that religion, with all its unreason, is the only show in town. The approach advocated here is therefore not to sublimate the phenomenology of our emotional needs as some separatists would, but neither is it to soften on intellectual needs, whose rational and methodological requirements are laid out in Chapter 5. We wish, in other words, to embrace both aspects of our spiritual quest and celebrate their union in an entirely naturalistic setting without any recourse to supernatural divinity. From that union we will try to reframe the problem as a quest to understand a comprehensive and quintessentially human trait: the search for beauty.

The arguments to be used in making this case rely on ideas from evolutionary biology and neuroscience, both experimental and theoretical. Some of the neuroscience is still in the active debate stage of scientific inquiry, so it would be fair to say that the overall presentation here is a hypothesis. But it is, in fact, a scientific hypothesis, not a just-so story or a mystical-magical what-if. These ideas are both testable and falsifiable.

As a guide to help place the forthcoming material into a framework, a short précis of the two main arguments in this chapter are given here. The first argument introduces the *prototype theory* of aesthetics, and it goes like this:

1. Evolution made familiar situations rewarding because they're safe.
2. Prototypes seem familiar because of the way memory works.
3. Therefore, prototypes are rewarding.
4. Experiments show that prototypes are beautiful – now we know why.

The second argument is a hypothesis for the neural origin of prototype theory, and it goes like this:

1. Neural oscillations, or *rhythms*, underlie stimulus representation and familiarity judgments.
2. Serotonin, neural rhythms, and reward seem to be closely related.
3. Prototypical stimuli might produce stronger neural rhythms.
4. Prototype-induced beauty produces reward through neural rhythms and serotonin release.

You shouldn't understand all these steps or believe these claims yet; this is just a preview to help place the material into context.

Evolutionary pressures

We try to make sense of the world in various ways, and when we succeed, we benefit. For example, we try to categorize, classify, and predict just about everything we encounter. Most of this behavior is unconscious and precognitive. For example, what do you see in Figure 1-1?

Most likely, you did not say "a number of patches of dark gray, with long slashes of spatially correlated lighter gray" or "a collection of black ellipses surrounded by..." You simply said, "a lion." It is in fact very difficult to look at this image and not "know" that you are looking at a lion. Even more fundamentally, it is very hard to look at this image and not know that whatever "it" is, lion or not, that it is some kind of animal sitting on a rock, surrounded by grass. Animal. Rock. Grass. These are all classification and categorization behaviors that betray your tendency to parse the world into comprehensible chunks.

Figure 1-1. What do you see here?

You might object that, all philosophizing aside, this actually *is* a picture of a lion. Fair enough, but the point here is that such a statement constitutes knowledge, and the world has not given you such knowledge. What the world has given you is photons, from which you gradually discerned that there are such things as lions in the first place. Only after that is done can you say what this picture "really" contains.

Why do you do this? Rather, what selective advantage accrues from doing it? Well, obviously, if you understand what you're looking at or experiencing, you can choose behaviors that lead to more desirable futures. What's more, evolution has built into you a circuit that makes you happy when you do this well: we are all familiar with the satisfaction of remembering a location, fixing something, or generally just being right. That satisfaction can range from mild enjoyment ("Yes, I thought that was a plantain") to the exultation that comes from a true "Eureka!" moment ("Omigod, that means that E=mc²''). These are all rather mundane observations of human psychology, but the underlying mechanism may be very important. This is, after all, a key aspect of how our minds work. Can we move beyond the fact *that* we feel this way and explain *why*, neurobiologically, we do so? What circuitry makes us search for good predictive theories? How do we evaluate them? In short, what neural mechanisms underlie the search for order?

Consider one of our ancestors, struggling to survive in the Pleistocene. As it travels through the world, raw information streams in through its senses. Regardless of how cautious it is, it must encounter new environments and unfamiliar stimuli all the time. Many of these things will be neutral with respect to survival – a strange rock or a piece of a leaf, a particular twinkle in a stream, or a rare tint of orange in the sky. In fact, the vast majority of what this creature experiences will be unfamiliar in its particulars. Certainly rocks and streams have been encountered before, but each individual rock is at least slightly different from the last.

While many of the unfamiliar experiences are neutral, others are beneficial. Perhaps this creature is sufficiently bold to take an exploratory bite of something, and gets rewarded with a nutritious lunch. Or perhaps it ventures into a dark cave and, finding it abandoned, commandeers a new home. These fortuitous experiences give it a leg up in the competition for survival and help ensure that it will survive and reproduce, passing on its genes to the next generation.

We should probably expect, however, that overzealous interaction with unfamiliar stimuli is in general a bad idea. Most things in the world will be neutral, many will be dangerous, and a few will be helpful. Just imagine you are standing in the middle of the forest: there are more ways to get hurt than to get fed. Even the mere act of walking around at random is costly, because it expends energy. As long as this is true, the expected payoff to an animal engaging in acts of spontaneous curiosity and exploration is low. This is especially so because the good things that happen at random are, in perspective, relatively minor. Free lunch or a new home are perks, but in order for them to substantially contribute to the animal's fitness, they must either have very high payoffs or happen consistently. In contrast, bad experiences can be very bad. Eating a poisonous plant or wandering into an occupied tunnel can cause sudden, severe, and irreversible harm. Most animals are more cautious than curious, and even curious ones – cats, raccoons, ferrets, otters – are careful when they explore, always ready to bolt if something goes wrong.

The alternative to extroverted curiosity, of course, is the way of the couch potato. This is a humdrum, predictable existence that shuns exploration and experimentation. With this lifestyle, just about everything encountered is old-hat. This may sound boring to an over-stimulated modern human, but try to see things from the more survivalist perspective of a hunter-gatherer. Only when local resources are so low or competition for mates or territory so high that the status quo is inherently threatening to fitness will striking out for a new niche pay off.[1] For all these reasons, equipping animals with a familiarity preference – some neural circuit that recognizes when things are familiar and activates reward centers in the brain – may give them a Darwinian advantage.

In addition to being safe, a familiarity preference yields another benefit. Knowing that something is familiar means that you are able to categorize it against prior experience, and if you're able to do that, you can make useful inferences. If you see a red, round, tennis-ball sized object hanging from a squat tree, you may draw on previous experience and determine that the object is an apple. In doing so, your knowledge of apples becomes available. From prior experience, you know that this thing you're looking at now – which you have not even tasted or smelled – is edible. It contains water and sugar. Perhaps you even know that you can plant the seeds and grow more apples. None of this information is available purely from the photons now striking your retina.

So. Recognizing an object allows inference of hidden attributes. This is also true for the recognition of familiar *situations.* Suppose that a gust of wind blows the apple off its tree. You have seen passive airborne objects before, and your memory of their paths allows you to predict the trajectory of this one. Or suppose you wander into elephant territory and witness the first stages of pachyderm irritation. From experience, you may be able to decode these behaviors and determine which are likely to lead to your flattening, which are bluffs, and so forth. All your stored experience helps you select the best possible course of action, but only to the extent that the object or situation is familiar.

On top of the direct benefits, there are second-order benefits controlled by your expectation of how much you will enjoy this feeling of familiarity. If you expect to like something more than your neighbor, you will probably do it more often and be better at it in the long run. This is as true for evolutionary selective pressures as it is for piano practice. The quest to find familiarity in initially unfamiliar settings could manifest as a sort of curiosity, but it is not the open-ended exploration of the vagabond. For the trait to survive selection pressure, its encouragement of familiarity-seeking behavior would have to pay off often enough to compensate for the initial overhead cost of entering into possibly risky situations. In other words, second-order familiarity seeking amounts to the development of *intuition* that such exploration is warranted. When the intuition is good enough, the animal will learn over time that such data-gathering engagements pay dividends later on.

In short, the rewards of familiarity preference are great, and the *expectation* of reward is also a powerful motivating force for learning, which produces more familiarity, which increases the expectation, and so on in a positive feedback loop. I will argue later in this chapter that these simple mechanisms could be used as general tools directing the mental processes of evolving animals. If certain kinds of activities lead to greater Darwinian fitness, great. But if the animal generating those activities can know on-line, more or less right away, that what it has just done is good, even better. For now, simply note that the evolutionary accident of connecting an advantageous behavior to a psychological reward mechanism is itself evolutionarily advantageous.

Top-down processing

Ok – you try to classify experience and when you succeed, you benefit. How does the brain classify experience? As a preface to answering that, let's consider what happens in the early stages of perception. We'll use vision as an example, but all the other senses do effectively the same thing.

Light comes into the eye through the lens, which focuses the image on the retina. The retina is composed of several discrete layers of cells, each layer of which has a circumscribed functional role. In the first layer are the photodetectors (rods and cones), which are responsible for actually transforming the electromagnetic energy of the incoming photons into electrochemical energy, in the form of ions moving across the membranes of neurons. The photodetectors are arranged quite densely on the retina, more densely at the center (the *fovea*) than at the periphery. This is why it is hard to identify shapes and colors in peripheral vision; when we want to inspect a stimulus closely, we look directly at it, which places the image on the high-resolution section.

Still within the retina, subsequent populations of cells assemble the raw information provided by the photodetectors and essentially calculate derived characteristics of the visual world. A single ganglion cell, for example, will take input from multiple photodetectors and become a "red spot detector." Because the connections between photodetectors and ganglion cells are hard-wired and the photodetectors don't move around, a single ganglion cell is forever locked into providing information about only a tiny portion of the visual field. There is no "big picture" cell in the eye – what began as a large, almost continuous image out in the world becomes immediately pixelated by the retina. Eventually, the output of millions of ganglion cells is bundled together into a single cable, the optic nerve, and sent through the skull and into the brain itself. From that point on, all the brain knows about the visual world is contained in the activity of a large but finite collection of ganglion cells.

Because of the way that sensory information is atomized right at the beginning of neural processing by the eye itself, it's clear that if there is ever going to be a search for large-scale (bigger than the scope of a single ganglion cell) features, it needs to happen in subsequent layers of neurons. Only after those layers have done some additional work can the re-integrated stimulus be compared to memory and identified. For example, if you want to know if a line is continuous over a relatively long stretch, no single ganglion cell can tell you. You are going to have to build some additional circuits that integrate the activity of many ganglion cells simultaneously.

In neurobiology, a distinction is made between "feed-forward" systems and "feedback" systems. A feed-forward system is one in which the input from the world is processed by some population of cells, which send their output to another population, which does yet more processing, passes its output onto another population, and so on. Each population acts purely on the information received by the layers earlier in the process. Everything is sequential. In contrast, a feed*back* system is one in which the neurons of one layer send their output both forward to the next layer *and backward* to the neurons in a previous layer, perhaps the one from which they received their input in the first place.

To process large stimuli like lions, there has to be a pretty strong feed-forward component to visual processing. Many of the component features of the lion (*e.g.* are the ears rounded or pointy?) can only be recognized when assembled from contributions of very many edge detectors, color detectors, and so on. If you were designing a brain, you might stop there and leave the system completely feed-forward. As a matter of fact, human engineers have spent a tremendous amount of time trying to do just this, and so far, have not done particularly well.

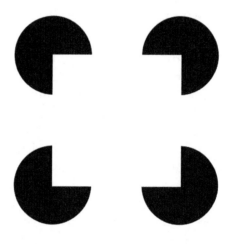

Figure 1-2. Kanizsa square

Nature, it turns out, "decided" to go the other way and put quite a lot of feedback into sensory information processing circuits. Unfortunately, the data sloshing that happens in feedback systems makes them much harder to understand and much harder to design. Even in systems like vision, where the feed-forward circuitry has been carefully studied, there are many recurrent connections from higher to lower layers. Such feedback signals are also called *top-down* signals. To understand the significance of these signals, let's consider another example. The lion is very real-world, but also pretty complex, so the purity of top-down effects gets lost. Consider the much simpler image shown in Figure 1-2, known as a Kanizsa square.

Do you interpret this image as a white square in front of four black circles? If so, you are experiencing top-down feedback from an early stage of the visual system, *i.e.* the one responsible for edge detection. In the world we know, most things are solid, spatially contiguous objects with continuous edges, so a white square in front of four black circles is a parsimonious and highly likely explanation. Of course, that's not what is "really" there. All that is "really" shown in the image, explicitly, is four black Pac-man like shapes.

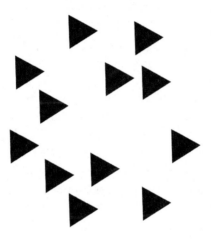

Figure 1-3. Scene coherence

In the next example (Fig. 1-3), all the triangles seem to be pointing in the same direction. There is no single "real" direction they share; it can shift from moment to moment. Northwest, southwest, and due east are all possibilities. But whichever one of those you "choose," every triangle in the scene adopts it simultaneously. As above, this coherence is not really in the image itself, but rather is forced upon it by your perceptual machinery.

The way in which neural networks mix feed-forward and feedback connections to compute is still largely unknown, but in a more general sense, the functional benefit of top-down influences might reflect the importance of classification. By implementing feedback, higher-level circuits are able to stack the jury, imposing their prior understanding of what is supposed to be there with what the senses are saying actually is there. This foisting of expectations onto perception is why it's so hard to look at the lion image and force yourself to just see a bunch of colors, and why it's hard to see neighboring triangles as pointing different directions.

The best way to interpret these results, then, is that higher-level systems try to interpret retinal imagery in a way that is true to the information provided *but also* in a way that makes sense given prior experience. Those higher-level systems have learned about the world and "know" that it is made up of objects (or just edges, or lions, depending on what level is involved). They "know" that objects have continuous boundaries, and things in flocks tend to move in a common direction. You can think of this as bias, but you can also think of it as a sort of a helpful suggestion. The higher-level system is pressured to parse its input as quickly as possible, and 90% of the time, probably more, such assumptions turn out to be correct. In some cases, the insistence that things make sense can cause problems and make you see things that aren't "really" there, but that's a story for another time.

In the visual system, this top-down influence is interesting and can produce some amusing effects, but there is a very important larger lesson. Not only do your senses filter the world, giving you access to just a small part of what's out there, but your cognitive machinery injects bias into lower-level sensory systems. In large part, you see what you know. But where does what you know come from?

Detecting familiarity

What you know comes from what you experience. It's possible that there's some built-in knowledge circuitry for certain coarse aspects of the world, but this kind of high-level cognition is so intricate that it seems unlikely that DNA could encode the synapse-by-synapse details necessary for, say, an *a priori* knowledge of lions. In general, you start infancy with an open mind, and each new experience updates your memory. Each new breed of dog that you see slightly alters your idea of the dog category.

In addition to updating memory, you also compare new experiences with memory to try and find the closest match. If you find a close but not perfect match, your brain will probably try to fill in any blanks based on more general knowledge of the category. If you had never seen a black poodle and then one day did so, your brain would probably make a quick inference based on size, shape, behavior, and so on, and decide that it was likely a dog. Once that connection is made, some of the attributes hidden in the current stimulus become available: you know it might bark, or bite. This retrieval process may begin crudely and become refined as more information becomes available, though, so it's possible something could potentially be *incorrectly* familiar. *I.e.* one could have the experience of recognizing something and then discovering that it was not, after all, what one thought it was. If, after you got a better view, the poodle seemed especially stocky, you might change your mind and decide it was a black bear. This example demonstrates our hopefulness in finding familiarity (you didn't remain totally ambivalent about the animal until you were positive, but rather, made a preliminary guess), as well as the fact that we can fail in the attempt. We *try* to find order in experience.

Advertisers are very well aware of the connection that this process has with our reward systems: raw familiarity makes people feel good. Marketers spend billions of dollars each year blanketing the world with logos, product names, color schemes, or whatever identifiable tags they have associated with their goods and services. If brand familiarity were not responsible for positive emotional connotation, then billboards, airport posters, and magazine spreads would have no effect on sales. Marketers have been at this for a while, and they know that indeed, just making potential customers aware of your brand, even if they don't know anything unique about your product or service, makes them more likely to buy it.

Of course, it's not just objects, in the sense of things you can *see*, that are familiar. Sounds, smells, tastes, movements, and types of situations, among other things, can also be familiar. Many of us have had the experience of buying an album of music and initially liking only a few songs. After mere exposure to the others, however, they grow on us and we come to wonder how it was that we didn't like them in the beginning. What's changed? Only the degree to which we're familiar with the stimuli. For you to perceive and recognize examples from each of these categories, there has to be some neural hardware capable of perceiving the stimulus, breaking it down, re-integrating it, and comparing it to memory, and of course there has to be a memory in the first place.

Now, if an animal is to be rewarded for recognizing when the world is familiar, there must be a purely neurobiological correlate of this occurrence. Of course *you*, as a self-aware being, know when something is familiar. The problem is – if this way of speaking can be forgiven – how your *brain* knows. The conscious feeling of familiarity follows the neurobiology; it would be tautological to use that feeling as the detection itself. So can we guess at what a neurobiological correlate of familiarity might look like?

Recalling the description of how the retina works, imagine the percept of an object – we can stick with the poodle as an example – as a collection of attributes or features. A certain amount of black here, a texture there, a particular style of walking. Some of these features go back to early stages of sensory processing, perhaps relayed straight through from the sense organ itself. A ganglion cell, for example, knows where colored spots are, and the next layer knows about line segments of particular orientations and locations. Again, the feed-forward construction of feature detectors in other modalities is similar to that in vision. Each feature is represented by the activity of a neuron or neurons, whose multiple activities are bound together into a single, coherent ensemble. Precisely how this happens is quite important and will be discussed below.

As this percept is measured up against others stored in memory, a component-wise comparison is done, determining the part-by-part overlap of each feature with the features of the object in memory. Perhaps the memory is of wider legs, or a shorter nose, or lower hips. Some neural circuit must determine the degree of match between the features of the current stimulus and those of the memory. The greater the similarity, the more this circuit is activated. Moreover, there must be a threshold in the similarity comparison, *i.e.* some line beyond which we decide that an experience is sufficiently like a memory that categorization is complete. Yes, it is a dog, or even better, a poodle. The finality of that decision is necessary for action. We cannot just sit in a tree and assign probability density distributions to a set of possible objects…in most cases we must decide, usually rather rapidly, which individual object we are dealing with and act accordingly.

Prototypicality

Familiar perceptions are safe and rewarding, and we have a bias to interpret the world through the lens of what we know. With the help of one additional piece, we will be ready to understand what this has to do with beauty. Religious experience will come a bit later.

In ancient Rome, the prevailing belief was that beauty was a property of the object itself, rather than in the eye of the beholder. Attempts were made to find mathematical constants in, for example, great architecture. The Roman writer Vitruvius gave, without justification, a value for the optimal ratio of column height to intercolumnar distance. More than

1500 years later, the application of ideal mathematical proportions to beautiful things was still extant in, for example, Leonardo da Vinci's celebrated Vitruvian man. And, of course, there is the cult of the golden ratio, which claims that the admittedly interesting mathematical constant $\dfrac{1+\sqrt{5}}{2}$ is nearly ubiquitous in nature and art. Behind each of these attempts, there are often interesting discoveries about human perception, some nice applied mathematics, and a sizable helping of self-fulfilling prophecy. If someone claims that the golden ratio is the best shape for a building, an architect might believe it. Then that architect goes out and erects a building with those proportions, and a third person points out how the golden ratio can be found everywhere!

Some of the work on analytical aesthetics must therefore be taken with a grain of salt. There is one modern result that I want to explore in some detail, however, because it is (a) empirically well supported and (b) a very productive platform for additional work. Additionally, the result has an interesting historical origin: it was first obtained by Sir Francis Galton, father of the now-infamous field of eugenics and first cousin to Charles Darwin.

In 1878, Galton was interested in studying the physiognomy of criminals. He wanted to know if there was some identifying facial characteristic that held true for "the criminal personality," which he assumed existed. This idea might be seen as a descendent of phrenology, which claimed that personality could be judged by examining the shape of the skull: the skull mirrors the shape of the brain and the brain's functions are localized. In any case, Galton wanted to find the shared facial characteristics of the criminal type, and to do this he needed a way to subtract out the irrelevant features that vary from person to person. The solution he settled upon was the creation of a composite photograph. In this procedure, a series of ordinary photographs of male convicts was taken. Each photo was controlled for size, lighting, and head position. Using that series, a new photograph was developed by using partial exposures: with ten originals, the shutter would be opened for one tenth the normal time on each component picture, and the composite would slowly emerge.

The procedure was successful and generated a new photograph that fully appeared to be a living person, though it was no such thing. Immediately upon viewing the result, Galton noted that the non-existent man in the composite was "much better looking" than the component men from which he was assembled. Galton's explanation for this rise in composite attractiveness was that irregularities of the component faces were averaged out. No two people are likely to have a blemish in exactly the same place, so over ten (or however many) exposures, skin tone and complexion average out to the point of invisibility. Also, symmetry was increased, because one man might have a droopy left eye, but just as likely, the next might have a droopy *right* eye. As plausible and straightforward as this seems, however, much subtlety lurks beneath the surface. The first is this: Which aspect of the composite is actually causing the rise in attractiveness? As I said above, the composite not only produces a face with average proportions, but it also smoothes skin tone and increases symmetry. We should somehow control for this. At the end of the day, Galton was right, but even he did not appreciate the complete import of his suggestion that average faces are more attractive.

Correcting for symmetry is fairly easy. All we need do is take a photograph of a real person (not a composite) and cut the photo in half vertically. Then, turn each half into a whole by making a left-right mirror copy and taping the two halves together. This yields two new "child" faces: one is a symmetric version of the original left half of the face, the other a symmetric version of the original right half. And the question is, are these perfectly symmetric "chimera" faces more attractive than the original? No, they're not. If you repeat the procedure for a number of original faces and ask subjects to rate both the originals and the symmetric chimeras for attractiveness, there is no significant difference. Galton didn't do this but his experiment has been replicated and extended by modern psychologists. Symmetry is not sufficient for facial attractiveness. First subtlety resolved.

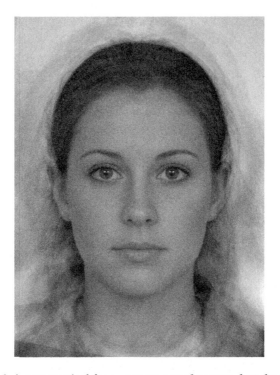

Figure 1-4. A prototypical face – not an actual person, but the result of morphing an image to match the prototypical geometry extracted from a set of 15 real faces.

To control for the skin tone / blemish issue, we turn to the Hollywood technology known as morphing.[2] This is when you digitally stretch an image and the computer interpolates to fill in any areas that expand, or compresses to eliminate areas that shrink. To do the morphing procedure, you need a set of keypoints. These are (x,y) coordinates aligned on every face, such as pupil locations, mouth edges, and so on. From that dataset, you can compute the *average* location of each keypoint: the most prototypical facial geometry. Then, you take any one of the component images and morph it so that its facial geometry matches the average. No smoothing. You get realistic skin tone without the soft focus artifact produced by photographic compositing. And? The result is that the average is *still* more attractive than the component faces. You can also do the converse experiment where you take a bunch of pictures of the same person and average them together. The slight differences in position from picture to picture mean that the same photographic averaging that happens with multi-person composites happens here, and the skin is correspondingly smoothed, but the morphed versions are still more attractive.

Thus, neither skin tone nor symmetry explain the rise in attractiveness that we observe with prototypical composite faces. Instead, it is the average *geometry* of the composite that is attractive. The eyes of a composite are neither exceptionally close together nor far apart, the lips neither exceptionally thick nor thin, the nose neither exceptionally wide nor narrow.

That sounds fine, but we all know that beautiful people are not common. Beautiful people are rare, yet this result seems to say that beauty is nothing more than averageness. The average person on the street is not beautiful, so what gives?

What gives is that you never meet the average person on the street. What you meet on the street are individuals, not averages. Essentially, this is semantic confusion over the word "average," which we use colloquially to mean "something that happens often." Yes, it's common that you see someone on the street, and they are only of typical, not exceptional, attractiveness. This is "average" in the sense that it is an everyday occurrence, but it is not average in the facial geometry sense. If you measured the distance between pupils, lip height, and nose width for a typical person on the street, it would not be the case that *every one of those values* would be very close to the population mean. You might get lucky and find, for a particular face, one or two of those numbers hovering around the mean, but most probably they would not all do so. Further, your brain tends to focus on the eccentric qualities of a face rather than the typical features, because it is precisely the uniqueness of a face that makes it identifiable. When you are confronted with a geometrically averaged composite face, which is average in *all* its values, you will rate it as more attractive than the component faces that make it up.

Another common observation regarding this result is that those faces widely acknowledged to be beautiful – supermodels, for example – are not average. The answer to this has an easy part and a hard part. The easy part is: Are you sure? The claim is generally made off the cuff, but to be rigorous, we'd need to do the calculation where the facial geometry of supermodels is calculated and compared to the composite geometry. If supermodel faces tend to be closer overall to the average than non-supermodel faces, then the substance of the complaint is dissolved. If not, then we have more work to do. As far as I know, the calculation has not been done, but my expectation is that it would turn out that the faces of supermodels are closer to the population average than the faces of non-supermodels.

The hard part of the answer deals with the inherent specialness of faces. We spend *a lot* of time looking at human faces, and we almost certainly have some hard-wired circuitry to process the subtleties of what we observe. Often unconsciously, we use visual cues to make inferences about the person we're looking at. For example: if the whites of the eyes are clear, we take this as a sign of health. This is why women wear eyeliner and eye shadow – the dark color contrasts with the eye and makes the white look whiter. Lip and skin color may be used to infer quality of circulation, and thus also indicate health – this is why women wear lipstick and blush. In men, secondary sexual characteristics like angularity of the jaw and facial hair indicate testosterone level, which in archaic environments (and perhaps modern ones as well) could be an important factor in mate selection. This may be why "the stubble look" is a sign of youthful virility, although there are certainly culturally conditioned aspects to that preference. Moreover, there are private personal preferences for all of these traits, and one can certainly find those who dislike makeup or stubble. The point is that the mere existence and the sheer number of sexually selected characteristics of the face confuse the issue of averageness as a basis for judging beauty.

Luckily, there is a way out of the confusion. All we have to do is pick some other stimulus for which there is no obvious sexually selected or otherwise Darwinian interference and do a similar experiment. Whitfield and Slatter did this with a stimulus category that, almost by definition, could not have any complicating evolutionary cues. The stimuli were pieces of furniture.

A collection of images were selected from three furniture design styles: Georgian, Art Nouveau, and Modern. In a preliminary experiment, subjects were shown three pieces of furniture from one of the styles, and asked to select pieces from the rest of the collection that were in the same style as the three target pieces. The preliminary results indicated that the Modern and Georgian styles were clearly differentiable, but that Art Nouveau was confusing. Many people considered Art Nouveau pieces to be merely bad examples of Georgian furniture. A few considered them bad examples of Modern furniture. Either way, Art Nouveau was not a clearly defined style of its own. With that understanding of subjects' categorization judgments established, a second experiment was done to measure aesthetic judgment. The setup was almost identical, with the three target items from a single style, but this time, subjects were asked to select items from the pool that, in their opinion, belonged in a living room with the three target items. By putting the emphasis on room decoration, the task became implicitly one about aesthetic choice. The results of this second experiment showed that subjects preferred prototypical pieces for each category. Modern went with modern, but Georgian went with *either* Georgian or Art Nouveau. Art Nouveau was usually not even selected to go with Art Nouveau! The reason for this goes back to the preliminary categorization work. Subjects thought that Art Nouveau was merely bad Georgian...*real Georgian*, *i.e.* more prototypical Georgian, was aesthetically preferable.

To connect this furniture result to that for faces, one need only notice that prototypes and averages are essentially the same thing. A prototypical Georgian chair is an average Georgian chair. An average face is a prototype of all faces. In fact, this was precisely what Galton was looking for: the prototypical criminal face. This same result has since been replicated for color selection tasks (people prefer "average" reds) and in aesthetic judgments of music. *Prototypes are beautiful.*

What are we to believe about the undying piece of folk wisdom that "Beauty is in the eye of the beholder?" Simply this: prototype = beautiful means that prototypes are *sufficient* for aesthetic preference, not *necessary*. There is still plenty of room for person-to-person variability, not least because different people have different experiences, and thus prototypes must be calculated within the individual. Beyond that, one person may, for reasons we may never discover, find fair skin attractive, another may like dark eyes. Such individual preferences can be shaped by experience, including that influenced by culture. Prototype theory certainly does not exclude such bases for aesthetics. What it does show is that a core component of the machinery underlying the aesthetic response is innate.

Finally, it should be obvious that although prototypes are beautiful, they are not *all* that is beautiful. There are other routes to aesthetic satisfaction...and other sources of pleasure, like eating and nice weather. This is fine; we are not trying to formulate a theory of happiness. There are many Darwinian forces that influence the reward system. Beauty is an interesting one because it represents a highly developed human quest and because, as we will see, it lies at the root of so much idiosyncratic psychology.

Cortical binding and neural rhythmicity

So far, this chapter's argument has been that familiar stimuli are evolutionarily advantageous, and it is the apparent familiarity of prototypes that makes them attractive / rewarding. To make that argument more solid, we need to move beyond the evolutionary motivation for making beauty pleasurable and find a *mechanism* that explains how the brain makes it happen. Because of the connection between prototypes and familiarity, the way we're going to do this is by investigating the way that brains (a) represent stimuli, (b) determine how prototypical or familiar they are, and (c) activate reward systems that make us feel pleasure. Those are the three remaining goals for this chapter. By the end, we will have concluded our investigation of why we search for beauty, and then, in the next chapter, we'll explore how that search manifests as science, religion, and more. For the moment, let's try to understand how stimuli are represented.

As mentioned above, a natural way to think about representations of objects is to imagine them as collections of features. For a visual stimulus, low-level features could be edges, colors, textures, shininess, *etc.* Using such elements as building blocks, you can then construct more complex stimuli out of pieces with various spatial or temporal relationships. In just about every type of sensory cortex, low- and high-level feature detectors have been found.

In principle, single neurons could represent complex objects (and this appears to be done in some cases), but the prevailing view is that stimuli are represented – in general, not just in visual cortex – by *populations* of neurons. These populations consist of all the activated feature detectors, bound together somehow so that downstream processing areas know which components should be grouped with which. This gluing together of the pieces to form a whole is known as *cortical binding*. Remember, too, that the sensory organs disassemble all stimuli no matter what. It's not like there's an option to just keep the original object intact, as a whole, all the way through the perceptual process. Objects are broken down, feature detectors are activated for isolated parts of the object, and the activities of those feature detectors must then be reassembled to perceive the complete object.

It turns out that there is a very interesting difficulty associated with cortical binding. Not surprisingly, it's called *the binding problem*, and it goes like this: if each complex object in the environment activates large sets of cells simultaneously, how will the brain know where the representation for one object ends and another begins? For example, when we're presented with a red square and a red circle, three feature detectors activate: red, square, and circle (see Fig. 1-5 for an illustration). Downstream neurons observing activity in these three detectors obtain a more-or-less clear interpretation of the world: There's no way to mislabel a shape with a color, because only one color detector is active. But suppose instead that we see a red square and a *blue* circle. Now, four feature detectors are activated: red, square, blue, and circle. If these detectors are all simultaneously active, the downstream interpretation is ambiguous. Which shape is which color?

There are several ways the binding problem could be solved by the brain. One way is *binding by convergence*. In this approach, all the component features of an object send their output to a single neuron which, when activated, represents the presence of the compound stimulus. In this case, you would need one neuron that received input from the red and square detectors, and only fired when both of the presynaptic cells fired.[3] You would also need a blue circle detector and, even though these shapes don't appear in this example, a blue square and red circle detector to be used in other situations. Only the number of incoming features limits the complexity of the stimuli that such anatomical convergence could represent. For a large fan-in, *i.e.* a large number of feature detectors converging on a single target neuron, a cell with a very specific preferred stimulus could be produced. For example, you might have a single neuron that fires only when you see your grandmother. In the early literature on this subject, this was exactly the example given, and such high-level feature detectors came to be known as "grandmother cells."

The binding by convergence proposal has some disadvantages, however. Think of the number of grandmother cells that would be required to represent the richness of the visual environment. For every possible combination of low-level features, there would have to be a new grandmother cell. There would have to be an unripe banana cell. A corkscrew cell. A French toast cell. A *stuffed* French toast cell. Considering the number of things we might possibly see, it is clear that we simply do not have sufficiently many neurons to represent all visual stimuli with grandmother cells. Moreover, the wiring necessary to accomplish so many converging and overlapping axons would far exceed the volume of our skulls.

As first suggested by Milner, and later clarified by von der Malsburg, a second possible solution to the binding problem is to replace convergence in space with convergence in time, *i.e.* have sets of feature detectors firing at the same moment. This is known as the *temporal correlation hypothesis*, and it's an important part of the overall hypothesis of this book.

The way this scheme works is that dynamic assemblies of neurons coordinate their activity temporarily to represent multi-featured stimuli. In the circle + square example above, the temporal correlation solution would be implemented by the red and square feature detectors firing in correlation with each other, but out of correlation with the blue and circle detectors. In turn, the blue and circle detectors would fire in correlation with each other, but not with red and square. It's the fact that the cells fire at the same time that indicates to other neural circuits that the features represented by those cells belong together.

If we watch the firing activity of a typical neuron over, say, a few seconds, we'll see that it is generally irregular. Typical cells don't fire with clocklike regularity, but rather,

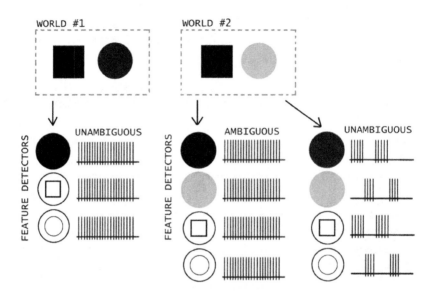

Figure 1-5. The binding problem and its solution via temporal correlation.

speed up and slow down, act noisy and staticky – more like a statistical response to the world around them than a perfectly predictable pattern. Even with that kind of irregularity, however, it's perfectly possible for two neurons to fire in correlation with each other. All that is needed is for the two firing patterns, however strange looking, to be similar. If the two patterns were identical, the correlation would be perfect. If a few spikes are shifted a few milliseconds one way or the other, or omitted completely, the correlation will still exist, only slowly diminishing as the firing patterns become less and less alike.

Given all that, it is perfectly possible that the brain could implement the temporal correlation solution without our really being able to detect it. After all, if a few or a few dozen neurons spread over the surface of the cortex fired in correlation with each other, how would we know? There is no way to know in advance which neurons we should record from, and even if we did, it's very difficult to get more than a small handful of electrodes going at the same time. I say that it is *possible* that the brain could do this, but in fact, it does not. In fact, the brain *does* use the temporal correlation solution, and luckily for us, it does so in a way that makes it relatively easy to detect.

Almost every time we find a set of neurons firing in temporal correlation, those neurons fire *rhythmically*, in short tight bursts of activity. The reason that this rhythmicity (at frequencies in the so-called "gamma" range of about 40 Hz) so often accompanies temporal correlation is only partially understood, but it makes detections of temporal correlation much easier.[4] Rather than just a few neurons participating in a binding circuit, many hundreds of neurons do so, and that group action results in electrical currents that are much larger and easier to record. Moreover, the oscillatory nature of these signals is easy to quantify. Whatever the ultimate reason for that rhythmic activity, it seems to be an exceedingly common if not universal concomitant of binding by temporal correlation.

In the spirit of disclosure, I should say that although the temporal correlation hypothesis and the rhythmic activity that goes with it now have very strong experimental support from a large number of independent researchers, it still is an actively debated subject in neuroscience. No one argues whether it happens at all. It certainly does. The debate is over how central it is to sensory binding. I emphasize it because it, particularly its rhythmic character, is essential to the construction of my main argument, as we'll see in a moment.

Before moving on, a brief recap. We are trying to understand how the brain makes familiarity rewarding, because familiarity is somehow connected to prototypes, prototypes are connected to beauty, and beauty is connected to everything. We have just explored how arbitrary stimuli are represented and learned that rhythmically firing neurons may be a key part of the picture. This fits into our overall quest to understand aesthetic experience in two ways. First, we want to demonstrate that naturalistic mechanisms, or at least hypotheses of those mechanisms, are capable of explaining "interesting" mental states. Near the end of the next chapter, we'll extend this description of aesthetic experience into the religious realm, but for now it is enough to see that we've taken a step forward from low-level reductionist ideas to high-level psychological ones. The second important thing that we've learned is that rhythm, as implemented by the temporal correlation solution to the binding problem, seems to play an important role in the representation of multi-featured stimuli. We'll ultimately be using that *neural-level* rhythmicity to tie in the *behavioral-level* rhythms that are sometimes associated with religious activity, in the form of rituals, chanting, dancing, drumming, and so on. Again, this is something we'll be covering in the next chapter. For now, just remember that (a) it is possible to explain high-level psychological phenomena, including certain aspects of aesthetic judgment, by making reference to low-level neural circuitry, and (b) there is something important about neural rhythmicity. With those points established, we can return to the question of how prototypicality is calculated, and from there, how reward is brought into the picture.

It is generally understood that memories are stored by modifying the strengths of synapses. The same neurons responsible for perceiving stimuli are capable of storing what is learned through such perception (à la Hopfield). Each time two neurons are activated simultaneously, the strength of the connection between them grows (this is a famous proposal by Hebb that has been experimentally corroborated by many, notably Bliss & Lømo and Kandel). If the activity of two neurons is anti-correlated, *i.e.* one cell increasing its firing rate at the same time that the other is decreasing, then the synapse becomes weaker. After some amount of experience, the values of the thousands of synapses within a network implicitly define one or more so-called "stable states" of activity, *i.e.* patterns of neural firing that "resonate" with the synaptic strengths. Because the weights have been influenced by many stimuli, they encode the average correlation between cells across all stimuli, and it is therefore precisely the stable states that represent the sense memory for prototypes. In general, the individual experiences will not be easily recalled; what is recalled is the prototype. The prototype defines the memory.

When a new stimulus appears, it induces some activity in the network where the stable state corresponding to the prototype is stored. If the stimulus is very close to the prototype, it is easy to imagine that rhythmically bound ensembles will be established quickly. Because the activity imposed from without is consistent with the stable dynamics indicated by the memories stored in the synapses, the dynamics settle down probably in about 200 milliseconds. Also – for stimuli close to the prototype – the strength of the oscillations will be greater than it would have been if the stimulus had been less prototypical.[5]

The calculation for how prototypical a stimulus is, then, is rather straightforward. All the brain needs to know is how quickly a stable rhythmic state is achieved and how strong the oscillations are. The more quickly it is established and the stronger the rhythm, the more prototypical the stimulus. This is consistent with cognitive experiments demonstrating that canonical, *i.e.* prototypical stimuli are recognized more quickly.[6] For example, the three-quarter view of a horse from typical eye height is easier to recognize than a view from directly overhead. Although the average viewer may have seen a few horses from directly above, the more canonical view is oblique and at eye level. Such views produce a set of synaptic weights that make that image a stable state in the neural network, so new horses seen from that perspective are recognized more rapidly (which is evolutionarily advantageous) and with greater confidence, and therefore – according to my thesis – seen as more aesthetically pleasing.

All of this brings us to the final piece: How is it that prototypical stimuli come to activate the reward system, thereby creating what is interpreted as aesthetic pleasure? I have already claimed that prototypical stimuli produce strongly rhythmic activity, so this question can be rephrased as "Is there a link between rhythmic neural activity and reward?"

Luckily, quite a bit is known about the neural mechanisms of reward itself. The brain areas responsible (*e.g.* the nucleus accumbens and amygdala) are well studied and the relevant neurochemistry (*e.g.* dopamine) also has a very solid literature. One of most reliable findings that has emerged from this work is the biological basis of raw, unprocessed pleasure, which turns out to involve release of dopamine in the nucleus accumbens. Drugs of abuse, especially opiates, are very effective at increasing the release, longevity, and/or effectiveness of dopamine in the accumbens. In fact, if you find a drug that causes dopamine release in the nucleus accumbens, or facilitates its effects there in any way, you can be pretty sure that the drug will be addictive.

Of course, the neural reward system wasn't created by evolution for the purpose of being abused. This same biology is the basis for the pleasure we feel when we do quite ordinary things that make us happy, such as eating, resting, having sex, and so on. Not feeling reward during such activities would lead, by definition, to ambivalence, and not caring whether you eat, sleep, or reproduce would be evolutionarily catastrophic.

Because the literature on the binding problem and cortical rhythmicity is relatively new, however, there has not been any direct work addressing whether rhythmicity of the sort used in cortical binding and reward are causally linked. There is, however, a good deal of circumstantial evidence in favor of the idea, and almost all of that evidence points to one chemical – actually, a neurotransmitter closely related to dopamine – that acts like a mediator between rhythmicity and reward. That chemical is serotonin.

Serotonin

The first requirement for any would-be mediator between rhythmicity and reward should be that the mediating agent is associated with rhythmic neural activity. For cells that produce serotonin, this is well established. Although the firing rate for these neurons is generally correlated with nonspecific behavioral arousal, their discharge rate is so regular and periodic that they are generally assumed to play a role as some form of pacemaker for other brain regions. The condition under which they are most active is during rhythmic motor activities like licking, chewing, and running on a treadmill (the experiments have been done mostly in cats and rats), though it is not entirely clear which way the causality flows; are they stimulated *by* this activity, or are they the cause of it? Rhythmic motor activity is not necessarily the same thing as rhythmic activity in *sensory* cortex, of course, but there is also experimental evidence that electrical stimulation of the serotonergic system increases an animal's interest in repetitive sensory stimuli.[7]

The anatomy of serotonin-containing neurons is also quite interesting, and places them in a unique position to coordinate large-scale sensory integration. Compared with most other neurotransmitters, serotonergic cells have an almost unmatched connectivity, each one contacting approximately 500,000 cortical targets. Seen the other way, each cortical neuron receives approximately 200 incoming serotonergic contacts. Other than that going to the cortex, the next densest collection of serotonergic fibers is the projection to the emotional centers of the limbic system, including the nucleus accumbens. This pathway allows serotonin to gate the effects of dopamine there; dopamine is the primary agent of reward, and serotonin modulates it.

Thus, from structural, stimulus-specificity, and modulatory considerations, the serotonergic system is in a good position to coordinate rhythmic activity over broad reaches of cortex, as required for diffuse cortical binding, and to modulate the brain's reward centers. Interestingly, the strong cortical innervation is selectively larger in primates than in rats, standing out against the background of the general neurological similarity shared by all mammals. Primates would not seem to have fundamentally different baseline sensory or motor needs from rodents, so it is possible that this serotonergic difference is related to a more fundamental *cognitive* difference between the animals.

Some researchers (*e.g.* Jacobs) have suggested that serotonin's dual relationships with rhythmic activity and reward explain why we sometimes engage in repetitive activities such as knee bouncing, chair rocking, and head bobbing. Because these behaviors are rhythmic, they activate the serotonin system, which in turn excites pleasure centers. This is essentially a form of self-medication; the behaviors themselves provide no real value. They don't get us any closer to food, or reproduction, or anything that might be construed as having a Darwinian return. Instead, we seem to have stumbled on a way to co-opt a system designed for something else, greedily and emptily pursuing worthless behaviors solely for the serotonin-mediated reward they produce.

As is often the case in neuroscience, a study of neurological disorders can be very instructive in our quest to understand the functioning of the normal brain. Acute clinical depression, for example, is sometimes treated with rapid eye movement (REM) sleep deprivation and electroconvulsive therapy (ECT). The therapeutic effect that has been reported with REM deprivation may be due to the fact that serotonin-producing cells are characteristically silent during this phase of sleep. A reduction in REM means that the serotonergic cells are prevented from going silent, so serotonin levels stay elevated. ECT achieves a similar end through different means. During ECT, a grand mal seizure is intentionally induced. During such a seizure, much of the cortex becomes rhythmically entrained to a single frequency. If the model presented here is correct, and serotonergic cells are activated by rhythmic stimuli, then ECT's antidepressive effect might be easy to explain: the seizure stimulates the serotonergic cells, which facilitate dopamine's activities in the nucleus accumbens. Reward (a.k.a. anti-depression) is the end result.

Other lines of relevant data come from the study of obsessive-compulsive disorder (OCD), Gilles de la Tourette Syndrome (GTS), autism, and epilepsy. A common symptom of OCD, GTS, and autism is rhythmic motor activity. Patients with OCD and GTS report having to repeat an action over and over until they get it "just right." Aside from the overtly rhythmic nature of these repetitive acts, it seems possible that some model for the behavior is imposing particularly strict requirements for satisfaction. Suppose the act in question is the turning off of a light after leaving a room. The motor cortex has a stored program for this behavior, and it compares what is actually done with the memory prototype to determine if it has been carried out correctly. This is analogous to the comparison of a sensory stimulus with long-term memory, and the same prototypicality / reward links apply. If the action is very close to the model program, binding is swift and strong, rhythmic coherence is rapidly established, and reward results. It seems possible that patients with OCD and GTS have unusually stringent requirements in this respect and find it difficult to produce an action that satisfies the comparator. Most actions do not feel like they have been done correctly, causing feelings of unease. In support of this hypothesis, both OCD and GTS are often treated with serotonergic agonists, *i.e.* drugs that enhance serotonergic activity. This is the same class of drugs often used to treat patients with chronic depression or anxiety, and in fact, unmedicated OCD and GTS patients prevented from engaging in their ritualized behaviors become depressed. They have been forced by their unusual neurochemistry to take to an extreme the serotonergic self-medication methods that a normal person would access in the form of knee bouncing and chair rocking.

The tendency toward ritualized behavior is also present and often even stronger in autism. The stringency of prototypicality comparison is taken even further in this disorder than in OCD or GTS, however, producing strongly aversive reactions to situations where the external world displays unpredictability. Whereas in OCD and GTS the psychological pressure seems to revolve around self-generated actions that must satisfy some norm, in autism these pressures require the *entire world* to be stereotyped. The anxiety produced in an autistic person by displacing a piece of furniture in their room, for example, would seem to indicate that some model for the room has been established and become so deeply entrenched that any variability is interpreted as a highly unfamiliar and aversive situation.

In addition to that magnified difficulty, people with autism also sometimes have an even more explicitly rhythmic, possibly self-medicating symptom: rocking and hand-flapping. As with OCD and GTS, serotonergic agonists are the clinical treatment of choice, though in autism these drugs are most effective in reducing the repetitive behaviors and stereotypy pressures specifically.

A popularized though empirically rare aspect of autism (and even GTS) is the savant syndrome, characterized by exceptional ability in one narrow domain. A possible explanation is that elevated serotonin levels, otherwise undesirable, facilitate binding and pattern recognition skills. A corollary prediction, stemming from the ostensible connection between the elevated serotonin levels and cortical rhythmicity, is that there should be high comorbidity of autism and seizure disorders, which is the case. In some cases of GTS, in fact, the spectrum of symptoms is reminiscent of temporal lobe epilepsy (discussed at length in the next chapter).

One way in which modern humans take advantage of the rhythm / binding / reward system is with drugs, particularly drugs that affect serotonin. Both LSD and MDMA ("ecstasy") exaggerate serotonergic function and have psychological effects that are easily understood if the rhythmicity hypothesis (as I'll call it henceforth) is true. LSD users report hallucinations reminiscent of synesthesia, that strange perceptual experience where different senses become spuriously glued to one another. One might, for example, smell a color, or see a sound. LSD users also report feelings of enhanced pleasure with rhythmic activity, especially music, and intense insight (the spurious sense that a given stimulus is precisely categorizable). The subjective effects of MDMA are less hallucinatory and more euphorigenic, but there are similar effects regarding reactions to music.

Reconsider the model: serotonin is the link between oscillations underlying feature binding, percept-memory binding, and reward. We know that LSD and MDMA facilitate serotonin, and these psychological effects seem consistent with an enhancement of binding circuits; this is why it seems that there must be (rather, there would need to be, for the hypothesis to be correct) a positive feedback cycle.[8] There would need to be both a forward mechanism whereby serotonin is released when binding is strong, as well as a backward mechanism whereby serotonin release facilitates binding. Under that model, the introduction of LSD or MDMA into a brain mimics (LSD) or actually increases the levels of (MDMA) serotonin in the synapse, and features that do not actually belong to the same object get spuriously bound in cortex. Because higher serotonin levels are also associated with mood elevation, this is pleasurable. Finally, the connections between *rhythmic* neural activity and the pleasure center makes periodic stimuli more rewarding, hence the elevated appreciation for music. At "raves," the music is heavily rhythmic, the drug of choice is ecstasy, and there's a credo / codeword, "PLUR," that (almost) ties the culture to the neurobiology: Peace, Love, Unity, Respect.

We should make one final note on oddities of the artificial manipulation of serotonin. If the process that allows current stimuli to settle into a stable state of memory became artificially enhanced, what would result? In analogy with the sensory effects of LSD, we might expect some spurious mnemonic recall, with the psychological effect that the user would mistakenly believe that something novel was familiar. If this happened diffusely enough, over the whole range of perception, the subjective impression might be familiarity with the overall state of affairs, as if the entire experience had happened before. Electrically, the associated parts of the cortex would be in a state of abnormally high rhythmicity, *i.e.* a seizure, but not so large as to cause the loss of consciousness as with a *grand mal*. I would suggest that this might in fact be the basis for déjà vu, which would explain why LSD users and patients with certain seizure disorders report experiencing it.

Considering all these lines of evidence together, it seems reasonable to hypothesize that serotonin may be a neurochemical mediator of cortical binding, neural rhythmicity, and reward. What I am suggesting here is that this mediation may be a key link in an explanation of how prototypical stimuli acquire their aesthetic appeal.

From prototypes to symbols

Linking together associated features and comparing percepts with category prototypes is, of course, not the sum of all cognition. If we wish to understand the characteristically human style of thinking we must also understand our tendency to use *symbols*. Presumably many other animals can classify their sensory experiences, but the natural impulse to denote those classes with abstract and communicable symbols is very rare. As it turns out, prototypes and symbols are deeply connected, and this connection will have an important role to play when the time comes to discuss the emotional power of the sorts of symbolism exhibited in nearly every religious practice.

Imagine our Pleistocene ancestor again, sitting in a tree, thinking about lions. He's encountered them in the past, escaped their attacks, and is now reflecting upon these experiences and trying to devise some method of victory over them. To do this, he needs to play with some kind of internal model of the lion. Clearly his sensory cortex can represent lions, so why not just use that? Because if his internal representation of the model is exactly the same as the representation when a lion is actually seen, the model will be indistinguishable from reality. In other words, using the primary representation for model building will cause hallucinations. To prevent this, some intermediary placeholder, or *token*, is required which will be used in all reasoning, predicting, and model building in general. Token use will, in fact, permit model building in the first place and will be a major evolutionary advantage. Using models, our Pleistocene friend can internally rehearse possible actions and try to anticipate the consequences of those actions without taking any real risks. The system of tokens and token manipulation that evolution and learning have produced constitute *language*.

Note that linguistic tokens are not necessarily things that can be said. It is quite common to use non-verbal tokens for mental objects, as you might do when comparing two tastes. You can think of them as "this taste" and "that taste" – though you probably do not go to the actual trouble of doing so – but there is obviously a depth to each idea beyond its "thisness" or "thatness." Somehow, an idea without a name still has a mentally graspable handle. Whether the objects of cognition are named or unnamed, physical things or abstractions, tokens are required. Language includes all symbolic manipulation, not just that which can be verbalized, gestured, or written.

As the richness of the internal representational scheme grows, a dictionary of tokens expands to include more and more of the perceptual universe. Some tokens, like "lion," denote objects out there in the world (at least, they seem to. What they really refer to is *our idea* of the objects out there in the world. More on this in Chapter 4). Others, like "headache," denote internal states. We also understand logical relationships, and see how things change over time. We develop linguistic structures to represent situations like "It's bigger than a breadbox" and "Mary gave the basket to John." Thus, tokens (words) can be combined into strings (sentences) representing states of affairs and experiences, both actual and imagined, and used in communication.

Although tagging a sensory or cognitive state with a label may appear to be a benign, almost passive act, it is far from it. In fact, in the labeling of an object, we often implicitly assert that the object belongs in a group with some other objects we've encountered before. By labeling the red tennis-ball sized fruit an "apple," we are making an assertion. Is that assertion guaranteed to be true? Not at all. Words are, in fact, theories.

Putting all the pieces together, then, we have the following model. When we perceive or contemplate a multi-featured environment, rhythmic neural activity is produced because neurons in disparate cortical areas need to coordinate their activity and solve the binding problem. A red detector in one area fires in synchrony with a circle detector in another area, then both shut off. Then a blue detector fires with a square detector, then they shut off, and the whole thing repeats, on and off, on and off, and that's what produces the rhythm.

Once individual objects have been parsed into oscillatory / multiplexed neural ensembles, each one is compared to and possibly bound to the long-term memory version of that same object. This is the process required to retrieve whatever attributes aren't superficially apparent. That percept-memory binding, in turn, activates the serotonin system. The complex feed forward and feedback circuits of serotonergic cells mean that the system works in the other direction as well, *i.e.* if serotonin is facilitated – *e.g.* by a hallucinogen – sensory binding circuits and percept-memory binding are facilitated.

Implicit in this scheme is the idea that the stronger the percept-memory binding is, the more strongly the serotonergic system will be activated. In other words, if you are looking at something that slightly resembles an apple, except it's very small, and more purple than red, then you might not really be sure that it's an apple, and the serotonin system won't activate much. That means that the pleasure center won't light up very much, and you won't get as much of that glad-to-recognize-what-you're-looking-at feeling. If, in contrast, you're looking at something that is just a perfect, prototypical, red, shiny, round, caricature of an apple, then the percept-memory binding will be very strong, the serotonin system will light up strongly, and you will feel happy. This is precisely the proposal, then, for how the familiarity preference system is implemented in the brain. The closer something is to your prototypical memory, *i.e.* the more familiar it is, the happier you feel about perceiving it. This will make you seek out familiar things.

From an evolutionary perspective, successful theory making – and remember, carving out a category and giving it a label constitutes just such an act – is rewarding for the same reason that sex is rewarding: it is nature's way of encouraging a certain behavior. Small theories that work yield small rewards. If you successfully line up a billiards shot, find the breadcrumbs in the supermarket, or recognize a rufous-sided towhee, you feel pleasantly smug. If, on the other hand, you predict the existence of the neutrino or cure Tay-Sachs disease, you get quite a rush.

Because of the ties that the neurobiological aspects of this proposal have to the idiosyncratically human strengths in theory-making and symbolic cognition, it is tempting to hypothesize that the serotonergic links that make it all possible were the result of a species-defining mutation or series of mutations. The psychological and cognitive consequences of those mutations would have been that the mutant animals would have enjoyed the process of successful theory making and would, over time, sought out situations that helped them understand and predict the world. This behavior would have made them smarter, more capable, and more formidable, enhancing their survival and advantage in natural selection.

Several paleoanthropologists, notably Richard Klein, have suggested that around fifty thousand years ago, there was a sharp increase in human linguistic and cognitive skill.[9] It was this change, they claim, that produced the advantages allowing Cro Magnons to displace the Neandertal culture that had been firmly established in Europe for almost 200,000 years. It is also at this 50,000 year point that the first widespread evidence of art, in the form of cave paintings and the celebrated Venus figurines, and music, in the form of bone flutes, also appear. We might understand how linguistic and cognitive advances could have produced a new species or subspecies with greater communication and social coordination skills, and thus presented dangerous competition for the Neandertals. At the same time, the relationship between rhythm, symbolism, and reward might have allowed the Cro Magnons to commandeer the underlying serotonergic neural hardware and obtain pleasure from evolutionarily neutral repetitive stimulation like music and art. The primary target of this neurochemical circuit, theory making, is of course *very* valuable; only these non-theoretical vacantly rhythmic behaviors are riding for free. And quite naturally, because *we* are Cro Magnons, all these practices are alive and well today, though as we'll see in the next chapter, they have taken on some new forms.

CHAPTER 2

Roads to Order

While awake, the brain is endlessly busy parsing the world and segmenting experience to find order in its input stream. Rather than perceiving a collection of unrelated features, we perceive coherent objects. Rather than perceiving a collection of unrelated tail-wagging animals, we perceive dogs. In each of these cases, the finding of order is the objective. But what is *order*? We all have an informal idea of what it is, but we needn't stop there. There are actually several well-formulated quantification schemes to make it more rigorous.

In this chapter, we begin by discussing the two main approaches to the quantification of order, namely information theory and algorithmic complexity. We'll then use both of those ideas, plus everything we've done so far, to understand how the beauty of art is related to the beauty of science. By the end of the chapter we will have assembled all the pieces necessary to show how the whole theme of the book comes together. Art, science, and the transcendent religious experience are all manifestations of the same search for order and beauty, all driven by an evolutionary need to understand the world, and all controlled by the same neural circuitry.

Information theory

Claude Shannon invented information theory in 1948 during his work at AT&T's Bell Labs. His goal was to quantify how much information could be transmitted over a telephone line, but the formulation he devised was sufficiently general that it has since been applied to all forms of communication systems and is particularly powerful in its discussion of *uncertainty*. I follow many information theorists in saying that a rigorous mathematical understanding of *order* can be achieved by seeing it as a *removal* of uncertainty. For readers with mathematical training, quantitative aspects of the following example are presented in Appendix A. For non-mathematical readers, it will be sufficient to understand that there *is* a way to be rigorous about quantifying information and order, and that this method works by basing itself on measurable probabilities. It is the connection of probability theory to our prior discussion of familiarity that gives our neurobiological framework additional support.

The more surprising an event is, Shannon said, the more information we get when we observe it. Let's say you are a monkey who goes to a particular watering hole every day to drink. Each time you're there, you notice the other animals, allowing you to develop a sense for the various probabilities of seeing an antelope, a wildebeest, or a hippopotamus. Over several years of visiting the spot, your implicit estimates of these probabilities are probably quite good, and let's say that in general they're also quite high – in other words, most of the animals are there on most days.

Now on some random day you bring to the oasis your visiting monkey-in-law, who has never seen it before. Upon arrival, you notice immediately that none of the usual animals are there. This is very surprising, so your attention is fully engaged. In essence, the world has transmitted a lot of information to you by generating the event of animal non-appearance. But to your visiting monkey-in-law, who has never seen this watering hole, and may therefore have a different sense of the probabilities of animals appearing at some other watering hole, a different amount of information is transmitted. It's more useful, probably, to say that the information is in the eye of the receiver, not the sender, because information varies with measured probabilities, and those probabilities are functions of experience.

This example demonstrates several things. First, the mathematical framework invented by Shannon can help us understand how our attention is engaged or not engaged to certain features of the world. Second (per Appendix A), it shows us how to do this in a quantitative way, not just a qualitative way. Third, this example shows that information is relative to your understanding of the *a posteriori* probabilities of the symbols involved. What is very informative or surprising to one monkey may be less so to another. Finally, we can see that from an evolutionary perspective, information might be simultaneously *relevant* and *dangerous*. In other words, getting into a situation where you are processing a lot of information indicates that the world is doing some surprising things, and therefore that you may not understand what is going on. In this particular case, there is ostensibly some cause for the absence of the animals. Perhaps a dangerous predator is lurking nearby, or the water has been poisoned. Not knowing why the world is acting in a particular way is generally unsafe. On the other hand, it is probably very important that you engage your attention, so that you learn from this experience, become more able to predict the events around you, and appropriately adjust your estimates of the probabilities involved.

Now let's think about this from a neurobiology of reward perspective. If you were evolution, and you were tinkering with the connections between the "information calculator area" (if there is one) and the reward area, what connections would you make? Would it be rewarding or anti-rewarding to see something that deviated from your expectations? Off hand, it seems that it would be anti-rewarding, *i.e.* unpleasant, to receive a lot of information from the world.

On the other hand, there is a possible second-order effect. Consider two groups of monkeys. Group A just avoids anything unfamiliar (we'll call them "Republicans"). Group B (Democrats) manages to figure out that approaching something unfamiliar, while risky, pays dividends if they prove capable of understanding it, because then the scope of their predictive ability is greater, and they can adapt to a changing world and capitalize on new opportunities. Which group will fare better in the long term? That depends on how dangerous the world really is and how smart the Democrats really are.

As this example suggests, information theory has applications in many domains other than the telephone business. For our purposes, the application to neuroscience is fairly straightforward (more so even than the monkeys-at-watering-holes) and very relevant. Although neurons are made from very different materials than copper or silicon circuits, and work in very different ways, some of their behaviors make them very amenable to information theoretic formalism. For example, the most noteworthy electrical characteristic of neurons is their ability to generate all-or-nothing "action potentials." Each neuron has a threshold, and if an electrical input causes the membrane potential to exceed the threshold, boom! An action potential is generated. The discrete nature of these events is very reminiscent of the discrete ones and zeros of digital computers. Any particular bit is either 1 or 0. In any particular time slot, a neuron either does or does not generate an action potential. If it does not, no information is transmitted to other neurons. Action potentials (also called "spikes") are *the* carriers of neural information.

While it was long thought that neurons encoded information purely in the rate at which they generated such action potentials (the *rate code*), recent discoveries have indicated that the precise timing of a spike train can carry information about the stimulus that produced it. Indeed, several laboratories have used information theory not only to decrypt the neural code – *i.e.* figure out how a signal is transformed into a spike train – but also to quantify how many bits of information are transmitted per second by sensory neurons. This technique has been so successful that it has allowed the experimenters to basically speak "neuron-ese," *i.e.* to look at the timing of a neuron's action potentials and reconstruct the stimulus. Miller and colleagues reconstructed a sound stimulus by measuring the spiking activity of sensory interneurons in the cricket, and Bialek *et al.* reconstructed a visual stimulus by recording the activity of a photoreceptor in the eye of a fly.

Thought of in this way, sensory organs can be seen as data acquisition devices. It is quite accurate, and more than a little eerie, to look at someone and think of their eyes as something like little security cameras, directed by a neuronal controller somewhere to point at and capture images from portions of the world deemed interesting. Areas of high visual contrast, locations where a surprising noise is detected, or parts of an object that would facilitate categorization are all high priority for downstream cognitive processors trying to do their job. When they need more data, they send a control signal to the eye muscles, the little cameras move, the signal is encrypted in neuron-ese and sent on its way.

The main contrast with the security camera analogy, of course, is that there is no security guard watching the monitors. There are only more devices. Ultimately, the result of all these devices watching each other is a motor output, *i.e.* your behavior. The real miracle of phenomenology is that there is no end of the line. No one lives in the house; the house itself is alive.

What does all this have to do with prototypes and beauty? Using faces as an example, we can think of the various dimensions of facial variability as dimensions in a vector space, or "face space." The value for the distance between pupils is one coordinate, the value for the distance from the bridge to the tip of the nose is another, and so on. In fact it is

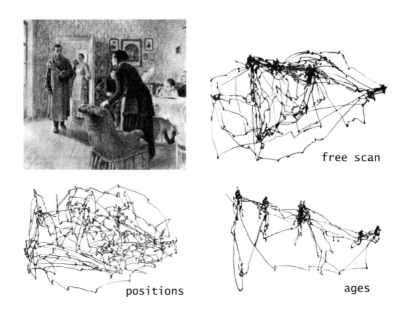

Figure 2-1. Visual scan paths. Subjects were asked to look at the image and prepare themselves to answer certain questions (people's ages or the positions of objects in the room) as indicated. Movement of subjects' pupils was tracked by a camera.

exactly these numerical quantities that the morphing software uses to construct the stimuli we considered earlier. If we collect data from a population of natural faces, quantify that data and plot it, we will probably observe something like a spherical cloud, albeit in this high-dimensional face space. Points near the center of the cloud correspond to faces with average values – a typical inter-pupil distance for instance – while points near the edges correspond to faces with more extreme features, such as a very high forehead. By counting the number of points within each small volume of the entire face space, we obtain an empirically-derived probability density that tells us how likely we are to find a randomly selected face in a given part of face space. Values near the center are more likely, values more eccentric are less so. By using that distribution, and the probabilities of individual "symbols" – which in this case are just faces – we can calculate how much information we obtain from any given face. And what do we find? Naturally, we find that average faces, which are very common, very prototypical, have very little "surprisal" associated with them. The statistical definition that averages are what we expect corresponds precisely to the idea that prototypes provide minimal amounts of information.

Plugging this idea into the spooky-but-true view of sensory organs as information gatherers, we obtain an even spookier but more explanatory one: the reason you can't stop looking at a beautiful face is because your brain is not getting very much information. Some neural mechanism sicced your little security cameras on the world and told them to gather information about this new face. The eyes turn and start to take measurements. How far apart are the eyes? How tall is the nose? All these quantities are calculated from the raw data coming in through the retina and processed by subsequent layers of neurons. If the face you're looking at is very beautiful, those processing centers will not be getting much information, because in the comparison with stored memories for faces, they (the layers of neurons) will discover that all the symbols have high probabilities and low Shannon information. The low information density is appealing but hard to believe. Look longer. Are you *sure* there's nothing else? Is the world *really* as safe as it appears?

What all of this tells us about the human cognitive / emotional project is that there is a quantitative framework for thinking about why prototypes are appealing. That framework builds a bridge between aesthetics and information. Prototypes are appealing – from a mathematical perspective – because they decrease the entropy of experience. Rather than seeing the world as composed of a near-infinite variety of things, we see it as composed of a much smaller number of categories, the prototypes of which are approached to a greater or lesser extent by any individual thing. With Shannon's help, this process can be quantified and studied scientifically, not just at the macroscopic level of category memberships but all the way down to the brain's most basic representations, *i.e.* neural spike trains. The study of cricket wind sensation and fly vision are great examples of how that analysis actually proceeds.

With these new tools, a new link has been added to the causal chain that ultimately will reach from Darwinism to God. From the story told so far, the chain now reads: Evolution, survival, recognition, familiarity, beauty, prototypes, information, neurons. (Don't worry – this chain won't get any longer, and in any case it's just a Hansel & Gretel way of keeping track of the path that brought us here).

Algorithmic complexity

Information theory, by virtue of the definition of Shannon entropy, allows us to calculate the unpredictability of a process. As we've seen, entropy and unpredictability are very closely related to information, and they are also closely related to the idea of *complexity*. We could sensibly define a string (of numbers, letters, or any kind of symbol) as complex if it takes a lot of information to convey it. These are all mature ideas that have been used for over 50 years in various branches of engineering and computer science, among other fields.

Whereas Shannon was concerned with characterizing the behavior of an ongoing process, a different but complementary way to approach the quantification of information is to look at just a single string, and try to quantify its individual complexity. This measure came later, arising from work in computer science. It is called *algorithmic complexity*, or *Komolgorov-Chaitin* (KC) complexity. The definition is simple: The KC complexity of a string is the length of a computer program needed to produce it. (As a matter of fact, it turns out that the KC complexity is very closely related both to Occam's razor – the general mandate to favor simpler explanations – and to Shannon entropy. This is reassuring because it reduces any sense of arbitrariness in our definitions, making it clear that complexity really is a property of the object being examined, not just a game we're playing with numbers).

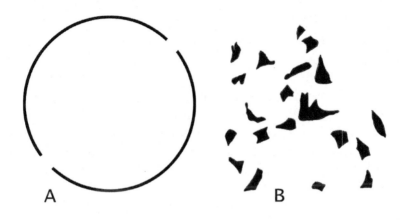

Figure 2-2. A. Continuation. B. Prägnanz.

To take a simple example, the string "01010101..." has a very low KC complexity, because it can be produced be a computer program that says, in effect, "Repeat the string '01' infinitely." The KC complexity of the marginally more complicated string "123123123..." has a computer program that is just one character longer. The decimal expansion for π, *i.e.* 3.14159265... has a yet larger KC complexity. And a truly random sequence of numbers has a KC complexity approximately equal to the length of the string because, by definition, there's no redundancy that would allow you to compress it. All you can do is repeat the string itself.

Algorithmic complexity also gives us another way to think about the perceptual consequences of top-down processing discussed earlier. In the world of our experience, it is more parsimonious to assume that the shape shown in Figure 2-2A is a circle than that it is two precisely aligned semicircles. It takes less information – a simpler program – to put a white occluding bar in front of a complete circle than to align the two semicircles so carefully. To take an even more powerful example, look at Figure 2-2B. Before continuing to read, do you see it as a coherent shape?

This figure is generally seen as a person – perhaps a knight – on horseback, even though the occluding whiteness would have to be a very unusual shape for that to be so. Is it "really" a man on a horse? Not really...there are just black splotches on the page. But as with the photograph of the lion in Chapter 1, the fact that those splotches *can be* interpreted as a person on a horse makes it almost impossible not to do so. The pressure to find simple explanations is very strong, and evolution evidently never saw much benefit in allowing us to consciously suppress it. Moreover, we are rewarded for the effort required to find order in chaos; if you didn't see the horseman originally but only after gazing for a few seconds, you probably experienced a pleasurable "Aha!" when the shape popped out. How might we explain this? From an information theoretic perspective, a disconnected collection of splotches is a high-entropy and therefore unrewarding stimulus. When an insight emerges that it is not something random, but something orderly – a figure on a horse – entropy decreases and the world becomes more predictable. Evolutionarily this is a favorable state, so we find it rewarding. How precisely does that happen at a neural level? In terms of the rhythmicity hypothesis, our initial look at the image activates a collection of unbound features detectors. Top-down processes try to force those isolated neural populations into something more coherent, and ultimately succeed in doing so by assuming the existence of a masking white foreground. This assumption allows some of the black curves and lines of the image to be interpreted as continuous, which is experientially a safe bet. When those neural populations are brought into the same binding circuit, the scope and power of neural oscillations increase. This activates the serotonergic system, which in turn activates the dopamine reward system, and we smile at the result.

Beauty is the pleasure you feel when a surprise makes sense.

Setting aside for a moment the question of whether KC complexity measures something objective, let us simply ask why it is that we *feel* like elegant, efficient explanations are more likely to be correct. That emotion comes from somewhere in the brain. Where? How?

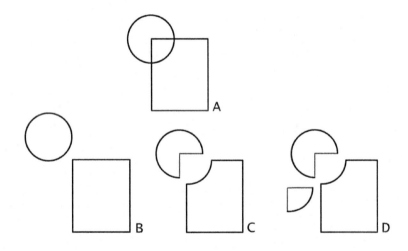

Figure 2-3. An image with multiple interpretations. The figure shown in part (A) could be a specially arranged view of the objects shown either B, C, or D. (B) is the most common interpretation because it is the most parsimonious, *i.e.* has lowest algorithmic complexity.

One possibility is that the emotional appeal of efficiency comes from the need to construct energy efficient motor plans. If we are sitting at the kitchen table and want to retrieve something from the refrigerator, our frontal cortex has to devise and execute a series of movements. In general, it will choose a sequence that has a minimum of spurious activity – we don't stand up and sit down twice, turn 360 degrees and walk backwards out the front door before returning to the refrigerator through the window. There are innumerable crazy ways of getting what we want, but only a very few efficient ways. Of course, those efficient ways require fewer calories than the crazy ways, and saving calories is evolutionarily advantageous.

Although some of the plans that humans create are far more complex than those produced by other species (as far as we know), efficient plan-making to *some* degree is not a uniquely human trait. Cats don't do ridiculous things on their way to the water bowl either. Even ants will generally not do spurious things while excavating, but this is probably because most of their action patterns are pre-programmed. In fact, some simple animals can be tricked into doing senseless things. Somewhere there is a movie of caterpillars daisy-chained into a rotating ring, each one following the tail of the next, all in motion, but with no net movement. This is probably possible because caterpillars lack the higher-order self-aware plan making hardware that would allow them to detect what is happening.

Humans are less likely than insects to fall into such pointless behavior, but only to the extent that our prefrontal cortex is doing its job well. The human frontal lobe, especially the forward-most part, the prefrontal cortex, is larger in proportion to the rest of the brain than in other species – indeed, you can use the degree of brain "frontalization" as an index along the evolutionary tree of human and proto-human species. It is also well known that patients with frontal lobe damage have difficulty planning, anticipating, and adjusting their plans and actions to new activities. One set of evidence for this comes from the Wisconsin Card Sorting test. This is a deck of cards, each of which has a unique image defined by three attributes: number, shape, and color. One card has, for instance, three red squares. Another has two blue stars. And so on. The person administering the test gives the entire deck to the subject and tells them to sort the cards based on one of the attributes – color, say. All the blue cards, regardless of shape or number, go into one pile. All the red cards go in another pile, also regardless of shape or color, and so on. The subject starts sorting. After a while, the experimenter changes the sorting criterion to something else – maybe shape. Now, all the cards with stars – any number of them, any color – are supposed to go into one pile, all the circle cards into another pile, *etc.* People with focal damage just to the frontal lobe can usually do the first part of the task but they have difficulty adjusting when the sorting criterion changes. This is called *perseveration* and is interpreted as a breakdown in the cognitive machinery responsible for constructing and generating rule-driven or goal-driven behavior. The same patients have trouble anticipating the emotional consequences of their actions, indicating that it is foreknowledge of behavioral consequences in general, not just something unique to the card-sorting test, that is affected by a frontal lobe lesion.

Just as a color or shape criterion on the Wisconsin card-sorting task helps our frontal cortex control the temporal evolution of world state through behavior, a theory (algorithm, heuristic) specifies how a system transforms its state from one moment to the next. Because primates, and humans most of all, have differentially larger frontal lobes than other mammals and also show the most well-developed planning skills, it seems plausible that the frontal cortex may act as a general-purpose forecasting machine. But that is exactly the sort of tool that would be required to determine the consequences of a theory. Because the frontal cortex already possesses an internal rating system that rates efficient motor sequences as emotionally appealing, it may rate efficient algorithmic sequences in the same way. The frontal cortical mechanisms that allow us to avoid the ring-of-caterpillars fate are the same ones that simplify our perceptual interpretations (Figs. 2-2 and 2-3) and make us prefer compact and aesthetic theories. One of the things that makes us distinctly human, in other words, is the frontal lobe driven quest to find beauty.

Prose, poetry and music

Another thing the frontal lobes control is sentence structure (= grammar) via circuits in Broca's area. If this part of cortex really is a forecasting engine, its control over the generation of allowable linguistic structures makes for an attractive picture. Just as an action plan requires individual movements to be put in a certain order to make sense, words must be put in a proper order to make a sentence intelligible. Evolution is conservative and tends to tweak rather than invent anew. If it had already created, for the purpose of action planning, a structure that can generate multi-step sequences constrained by the laws of body and world mechanics, why not plug that same recursive circuit into the communication machinery and use it to enhance, or even create, language?

Noam Chomsky and his intellectual heirs have argued that the apparatus of grammar is innate, *i.e.* not learned, and therefore shared by all humans. Although the superficial syntax of languages differs on issues like subject-verb-object order (English and French are SVO, Japanese and German are SOV), Chomsky showed that this difference was not deep. Careful analysis shows that all known languages, from Finnish to Swahili to American Sign Language, can be produced by a single *universal grammar* (UG) possessing just a few knobs, including the SVO/SOV distinction.

As we all know, there are many ways of saying the same thing. Given one semantic structure in UG, there are many possible corresponding speech acts. Consider:

- John threw the ball to Mary.
- The ball was thrown by John to Mary.
- Mary was the target of the ball thrown by John.
- John moved his arm and caused the ball to fly through the air to Mary.

These phrases represent the same state of affairs, but it is natural to call the first more efficient than the others. Efficient how? Certainly in the relative simplicity of its sentence structure, but more obviously in the number of words. This may very well be the origin of Strunk & White's compositional edicts we should all have learned in high school. Use the active voice. Omit needless words! The simplest way of saying something is usually the best. The source of these rules lies in the same neurobiology that we have already discussed. Namely, the frontal lobe circuits that control motor program generation and the rhythm- and serotonin-mediated preference for prototypes. Graceful sentences are the simplest ways to encode algorithmically the associated idea, and good word choices lie at the center of their semantic clouds. When John Greenleaf Whittier writes that "...Of all sad words of tongue or pen, the saddest are these: 'It might have been!'" he seems to capture *precisely* the gem of meaning, with nary an inessential syllable.

This is Komolgorov and Chaitin sipping serotonin tea with Occam and Shakespeare. The most powerful and appealing writing is usually the most concise, that which chooses the perfect words and puts them in the perfect order. The perfect words, of course, are those denoting the purest form of the expression, just as the perfect apple denotes the purest form of the fruit. Given language's ability to condense multiple forms of aesthetics into one while simultaneously delivering semantically discoverable meaning (unlike art, whose semantics are more obscure), it is not surprising that religion would find it more than a useful ally. Fundamentalist reverence for the exact wording of a holy text manifests in many faiths, from the torturous devotional letter-by-letter copying of the Judaic Torah, to the creation of entire schools of Arabic calligraphy that do nothing but exalt passages of the Koran, to the well-known Biblical passage (John 1:1) that "In the beginning was the Word, and the Word was with God, and the Word was God." Words and their aesthetic and religious content are so important that each of these faiths has come tantalizingly close to deifying religious language itself and dispensing with the God who is supposed to have said it! Mantra, anyone?

Like language, music possesses both a syntax (octaves, keys and chord progressions) and a semantics (the declamatory statements of a *forte* piano, the plaintive interrogations of an unresolved flute phrase). Indeed, the structures of instrumental composition often seem to echo those of the language spoken in the culture that produces it. The wild intricacy of tabla sounds somehow like fluidly polysyllabic Hindi; the tightly controlled Tsugaru shamisen sounds like punctuated and precise Japanese. Music also has internal structures of redundancy, in the form of rhythmic beat structures and mathematical ratios between the frequencies of notes. All of these traits provide platforms for pattern, aiding in the use of music as an elicitor of reward. One can imagine, for example, computing either a Shannon entropy or KC complexity for a musical score on a purely syntactic basis. Both the frequencies of notes and their timing can be measured objectively, so in principle there is nothing stopping us from doing that calculation. How the computed values related to aesthetic appeal, however, is probably not monotonic, because for must of us aesthetic preference lies somewhere between the dead simple and the impossibly complicated.

As suggested by Figure 2-4, most of us find perfectly predictable stimuli boring and completely unpredictable stimuli senseless. Musically and otherwise, the location of any given person's aesthetic sweet spot on that continuum is probably a function of experience. If you have listened to a lot of jazz and you understand its structure, you will probably enjoy it more than someone hearing it for the first time. Rhythmic music, as seems to be present in every culture, should be universally enjoyable, because of its special role as the common denominator of cortical binding. It is worth pointing out that even classical music, commonly contrasted with rock, also has repeating themes and canons that take

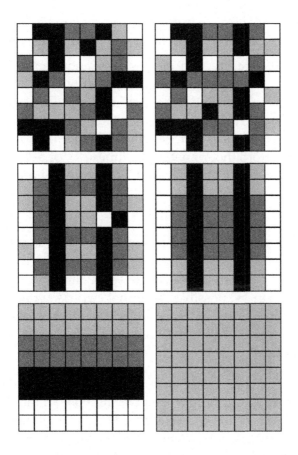

Figure 2-4. Grids with varying entropy. All but the one in the lower right are Scha grids, conserving equal numbers of each color across differing spatial arrangements.

advantage of the same neural prediction and familiarity circuits. One wonders if information theory could shed any light on whether the claimed sophistication of classical music has any empirical support or is based on mere classism.

In their book *The Rhythmic Structure of Music*, Cooper and Meyer describe some of the principles of musical composition: coherence of higher-level rhythms, groupings on higher levels, rhythm and tension, and recapitulation and illustration. Each of these principles can be interpreted in neuroaesthetic and cognitive terms: rhythm, entropy and algorithmic complexity, and the desire to find order in chaos.

At the intersection of language and beauty is poetry. By virtue of this it is poised as an especially good medium for expressing of the beauty of experience. As with prose, good poetry will satisfy Occam's razor and contain no unnecessary words – hence *le mot just*. It will probably also be rhythmic (with forms from the sonnet to the limerick), or at least contain repeating elements that facilitate the recognition / reward circuits.

*   *   *

The Chompskian argument relies not just on the universality of UG, but also on the remarkable speed with which children learn language, particularly in the sense that they seem to do so faster than is possible. In other words, children solve linguistic problems that seemingly cannot be solved solely through analysis of the language they have heard. As Steven Pinker explains:

> There is one more reason we should stand in awe of the simple act of learning a word. The logician W. V. O. Quine asks us to imagine a linguist studying a newly discovered tribe. A rabbit scurries by, and a native shouts, "Gavagai!" What does *gavagai* mean? Logically speaking, it needn't be "rabbit." It could refer to that particular rabbit (Flopsy, for example). It could mean any furry thing, any mammal, or any member of that species of rabbit (say, *Oryctolagus cuniculus*), or any member of that variety of that species (say, chinchilla rabbit). It could mean scurrying rabbit, scurrying thing, rabbit plus the ground it scurries upon, or scurrying in general. It could mean footprint-maker, or habitat for rabbit-fleas. It could mean the top half of a rabbit, or rabbit-meat-on-the-hoof, or possessor of at least one rabbit's foot. It could mean anything that is either a

rabbit or a Buick. It could mean collection of undetached rabbit parts, or "Lo! Rabbithood again!," or "It rabbiteth," analogous to "It raineth."

The problem is the same when the child is the linguist and the parents are the natives. Somehow a baby must intuit the correct meaning of the word and avoid the mind-boggling number of logically impeccable alternatives.[10]

From a neuroscientific perspective, the mechanisms of a would-be genetic specification of a UG circuit, which entails vastly more complexity than just word meanings, is even harder to understand. There is, however, a third possibility: UG may be extracted from *non-linguistic* experience.

It is perfectly obvious that each specific vocabulary item (the lexicon of a language) is acquired, not innate, regardless of any mentalese infrastructure built under purely genetic guidance. I argued above for the propriety of thinking of words as theories, because the mere act of categorization is an assertion. This interpretation makes sense for nouns and adjectives, even verbs and adverbs, but is a little less straightforward for other parts of speech, and less again for the algorithms by which words can be put together to form sentences (the grammar of the language). Your parents didn't point to something and say, "Look at the *sometimes* over there!" Strict Chomskyism claims that because UG is universal and learned suspiciously quickly, the neural support for it must be innate. But there is something other than genes shared by all language speakers, and that is the world. Whether we are raised in London or the Kalahari, we all experience, in broad strokes, the same natural laws with the same sensory apparatus. Just as we condense the static idea of Platonic applehood from a diverse collection of imperfect real apples, we distill theories of *dynamic* phenomena from the things we experience. Most of these theories are never given names or mathematical underpinnings, but they're there. If you saw the

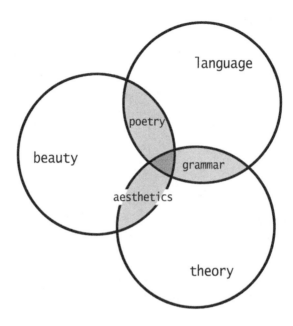

Figure 2-5. Venn diagram of beauty, theory, and language.

ball that John threw to Mary follow a rectangular trajectory, your implicit theory of ballistic motion would certainly be violated. Or, to take a deeper and more influential example, if you saw that ball launch itself into the air without John moving his arm, your implicit theory of cause and effect would be violated and probably prompt you to investigate. Even some deep theories are not present at birth. The expectation of object permanence, for example, does not appear until infants are eight or nine months old. Prior to that point, babies find the random and complete disappearance of material objects perfectly reasonable.

Just as a lexicon can be seen as emerging from the collection of static categorical theories extracted from experience, the framework of grammar could be interpreted as a distillation of dynamic theories, also extracted from experience. Subject and object, cause and effect, temporal dependencies, and many more *could* all be world-derived concepts that find their way into the frontal lobes, to be later applied to language and other forms of theories.

Note that I said *strict* Chomskyism. Modified versions of UG's origins are somewhat softer in the allowance for interplay between nature and nurture, and less committed to the innateness of a fully-developed, computer-like grammar machine. Pinker suggests that something as simple as an innate neural module for building recursive conceptual structures could account for much of the complexity of grammar.[11] This is reassuring from a neurobiological perspective because feedback loops are very common. Neural circuits tend to be rough-hewn, massively parallel sorts of affairs, which is hard to reconcile with the crisp precision of Chomsky's grammatical structures. He may be right anyway, but we should at least consider other options.

Although simplifying the world by (for example) regarding objects as members of a class entails information loss, reducing the entropy of experience is precisely what enhances survival. We may fret and complain that we are unable, in virtue of our pre-intentional linguistic habits, to represent the raw truths of the world, but this is the price we pay for the otherwise compelling advantages of symbolic representation. Language is useful *because* it is a lie.

Art

Although we certainly find beauty in natural things, such as faces and landscapes, aesthetics *per se* is most often talked about in connection with human constructions: in poetry and in prose, yes, but even more often, in art.

For Plato and Pythagoras, beauty was an objective, not a subjective property of objects, inherent in the thing itself. Prototype theory would seem to support that, at least in the sense that there are some aesthetic qualities that are not purely a matter of taste. Still, it seems unreasonable that prototypicality and parsimony effects could form the basis for all art appreciation, and I make no such claim. There are, however, other aesthetic principles, which can also be tied to human cognition and neurobiology, that can help us understand how the brain gives rise to the search for beauty. The first place to turn for a lesson in neurally-inspired art appreciation is, oddly enough, with rats.

We are going to train a rat to differentiate between a square and a rectangle. To do this, we'll present shapes on a computer monitor and reward the rat with food when it presses a lever during the rectangle's but not the square's appearance. After relatively few trials, the rat gets the idea that rectangularity is the signal to watch for, regardless of position, size, color, or anything else. We can measure the strength of the rat's certainty of this conclusion by measuring how vigorously (number of presses per minute) it hits the lever.

Now suppose that the rectangle we use during training has an aspect ratio of 2:1. The square is obviously 1:1. We can now ask how the strength of the rat's response changes if we alter the aspect ratio of the rectangle, *e.g.* make it longer and skinnier or more square-like. Presumably, as the aspect ratio nears the 1:1 value of the square, distinguishing between the two shapes becomes more difficult. We would expect to see, and we do in fact see, that the rat's confidence falters, marked by a corresponding decrease in the frequency of its lever presses. Conversely, as the rectangle becomes even longer and skinnier than it was during training – to an aspect ratio of 4:1, say – it becomes more and more easily distinguished from the square. Does this bolster the confidence of the rat and encourage it to respond more strongly to the 4:1 stimulus than to the 2:1 shape? Yes.

Niels Bohr said that a great truth is one whose opposite is also a great truth. Just as prototypes are effective elicitors of reward via familiarity, so too are their opposites, *i.e.* caricatures. This increased responding to an exaggerated rectangle is known as the *peak shift effect*, and it is easy to see that in less scientific contexts it is the same thing as caricature. The rat has brought us back to art. By exaggerating Jay Leno's chin or Nixon's eyes to impossible proportions, their portraits become easier to identify. For that matter, exaggeration of any distinguishing characteristic can be used to facilitate recognition. When a comedian impersonates a political or entertainment celebrity, they focus on those traits that distinguish the person from the average. Writing in 1844, Schopenhauer seems familiar with the practice and the reason for it:

> ...the arts whose aim is the representation of the Idea of man, have as their problem, not only beauty, the character of the species, but also the character of the individual, which is called, *par excellance*, character. But this is only the case in so far as this character is to be regarded, not as something accidental and quite peculiar to the man as a single individual, but as a side of the Idea of humanity which is specially apparent in this individual, and the representation of which is therefore of assistance in revealing this Idea.[12]

Cartoon caricatures are so common now that it is easy to see them as a modern phenomenon of illustration and mass media, but in fact their origin stretches back much further. Michelangelo portrays Adam as a God on the ceiling of the Sistene Chapel and David as a god in marble, but he was carrying on an artistic tradition from classical Greek and Roman sculpture. There, through exaggerations of size as well as sexually dimorphic traits like masculine angularity and feminine curvaceousness, features were enhanced to set the gods apart from ordinary mortals. Heroic busts, reliefs, and temples attest to the same tendency to isolate and magnify salient features. Though celebrated, even these are not the beginning. Four thousand years before the Greeks, Sumer's hero Gilgamesh was described like this:

> Supreme over other kings, lordly in appearance,
> he is the hero, born of Uruk, the goring wild bull.
> He walks out in front, the leader,
> and walks at the rear, trusted by his companions.
> Mighty net, protector of his people,
> raging flood-wave who destroys even walls of stone!

> *- Tablet I, The Epic of Gilgamesh (Kovacs, trans.)*

When a writer creates a character anemic in most real human traits but exemplary in one or two, like heroism, villainy, or greed, it may because the caricature helps the reader understand who (or rather, what) the character is. It is easier to create a grotesque than a real personality, which is probably why the practice of doing so is regarded by literary critics (and perhaps most readers) as a kind of cop-out, something too pat, too simple to be real. On the other hand, perhaps it is the need to establish what is a literal accounting of history and what is religious myth that gives godlike characters such immortality. Similar characterizations can be found in the hero sagas of just almost every culture, including 21st century America – witness Batman, Wolverine, and The Hulk.

Even Sumer, however, with its status as the origin of civilization, doesn't lay claim to the beginning of caricature. In fact, the real origins stretch all the way back into human evolutionary prehistory where, forty thousand years ago in the caves of paleolithic France, Cro Magnons painted antelopes, horses, and wooly rhinoceros that – were our lives still dependent on them – we would no doubt depict in much the same way. The bellies are absurdly plump, the feet and ankles tiny, the horns stretched and narrowed...all the identifying shapes are caricatured. For the hunter-gatherer culture of the Cro Magnons, the nutritional value and defensive hardware of a prey animal are obviously salient

Figure 2-6. Venus of Willendorf. Carved circa 23,000 BCE.

features. The famous Venus figurines, such as the Venus of Willendorf (Fig. 2-6), date a bit later, but seemingly share the same motivations. Taken to be fertility goddesses, these are female – actually superfemale – figurines. Hips and breasts are exaggerated, and body fat is higher than was probably achievable in that environment. Of course, these are sexually dimorphic features for women and relevant to successful childbearing, explaining their selection as the points of interest for caricature. Ramachandran and Hirstein have also emphasized the importance of the peak shift effect, especially as applied to sexual dimorphism, in sensual Hindu art. They further note neurological evidence for a link between categorization (which is facilitated by caricature) and reward:

In Capgras syndrome, for instance, connections from the visual 'face region' in the inferotemporal cortex to the amygdala (a part of the limbic system where activation leads to emotions) are severed so that a familiar face no longer evokes a warm fuzzy emotional response (Hirstein and Ramachandran, 1997). Remarkably, some Capgras patients are no longer able to link successive views of a person's face to create a more general perceptual category of that particular face. We suggested that in the absence of limbic activation – the 'glow' of recognition – there is no incentive for the brain to link successive views of a face, so that the patient treats a single person as several people. When we showed our Capgras patient DS different photos of the same person, he claimed that the photos were of different people, who merely resembled each other! One might predict, therefore, that patients like DS would also experience difficulty in appreciating the metaphorical nuances of art, but such a prediction is difficult to test.[13]

If one accepts the lesson of the peak shift effect, *i.e.* that caricature of highly salient features makes a stimulus more recognizable, then one gets for free an understanding of how caricature might yield reward in some form. Evolutionarily, recognizable situations are safe, safety should be rewarded, so caricature becomes rewarding – it's that simple. Caricature may not be *beautiful*, but the fact that we tend to produce it and look it means that it is rewarding (in this context, a behavior is rewarding if we tend to do it spontaneously). There remains, then, a need to reconcile this with the reward produced by prototypical stimuli. Prototypes are by definition averages, while caricatures are by definition eccentric. How can both produce reward via the same neural mechanism?

The answer, I think, is that reward produced by prototypicality is a reward for classification, while that produced by caricature is for individual identification. When you see a beautiful face, you enjoy it not because you know who it is, but rather, you feel as if you are gazing upon something eternal. (And, in a way, you are: you have probably seen sufficiently many faces to make the average relatively stationary). Having identified the object of your gaze as a face, the brain tries to further identify it as an individual.[14] It seeks out those features that distinguish it from the average. In the case of a true prototype, there are no such features, so the brain looks and looks and looks and fails to find anything. This search for individuating information is precisely what causes you to get visually stuck on beautiful faces – you just can't stop looking for more information. Your attention is captured, and your reward system is activated.

In the case of a caricature, reward will result as long as the features caricatured take the object further away from other category prototypes. If, for example, a cherry were caricatured by exaggerating its size, it might begin to resemble an apple. This would interfere with rapid and strong binding, so we might predict that subjects would find this less appealing than a normal cherry. (Anecdotally, I am fond of shrimp cocktail but downright terrified of six-inch prawns). This is why faces are emphasized and bodies de-emphasized for political cartoons. We tend not to notice or care about a vice president's legs, but for Tina Turner, those are salient features and would be caricatured.

While facial caricature is rarely attractive, this may be because the deformations introduced would indicate, in a real person, a severe genetic or developmental abnormality. In cases where exaggeration would indicate unusual evolutionary *fitness*, caricature will indeed be attractive – whence the aesthetic origin of breast implants, weightlifting, collagen injections, anti-wrinkle creams, and so on.

We have seen how prototypes form a basis for aesthetic judgments. Now it appears that caricature too can be

nt feature of the aesthetic, especially in the artistic
it seems clear enough that as a form of

**Figure 2-7. Landscape by Bierstadt. 19ᵗʰ century. A peak shift caricature of
a landscape.**

exaggeration, caricature is anything but prototypical. And yet
there is something about a caricature, of a person for instance,
that we seem favorably disposed to call a prototype of that
person. Norman Rockwell was a master at taking advantage of
this complementary relationship to convey the essence of a
scene. Consider, for example, his painting *Mysterious Malady*
(Fig. 2-8).

To see how this work uses caricature, think of eight or
ten attributes that might indicate a wholesome country boy.
Make a list, and rank them in order of importance. It's a safe
bet that at least your top five items are represented in this
image. It's not that the *way* they are represented is extreme –
Rockwell's boy doesn't have cheeks that are too freckled or his
pants too worn at the knees; it's the *number* of attributes that
are represented at all. In any real world situation, you don't
really expect every single one of the defining attributes of
some category to be present in every instance of that category.

When they are all present, you have a kind of "hyperrealism" or Platonic idealism. It is an extreme, but it is also a prototype. It is a representation of a large number of prototypical traits.

Norman Rockwell

Figure 2-8. Norman Rockwell's *Mysterious Malady.*

This is nearly a universal practice in Rockwell's paintings. Rockwell himself said that he was trying to "interpret the typical American."

Interestingly, it is difficult to make a grotesque caricature of a beautiful face. This was done experimentally with the facial morphing technique, but the result was that subjects found the caricature marginally *more* attractive than the original.

It has been said that art is about "snatching the eternal from the desperately fleeting" or "the seeking of knowledge in an ever-changing world."[15] I take a little license and interpret the eternal as the population average and the fleeting as the idiosyncrasy of an individual.

The relationship between prototypicality, caricature, and beauty, then, is this. When the stimulus at hand is prototypical of its category, the categorization machinery goes quickly and the rhythm / serotonin hardware yields feelings of reward. If, in addition, the stimulus calls for individual identification, the presence and exaggeration of unique traits make that recognition also proceed rapidly and yield reward in the same way. The two processes rely upon the same underlying circuitry but can activate it independently. A poem might perfectly capture the emotion of yearning, as a prototype, by highlighting the traits of a character who expresses the feeling. A musical score may use a minor key to exaggerate the discordant frustration of an unrealized goal, but fail to do so well enough to strongly activate our category prototype. In either case, the prototype or caricature facilitates our desire to clarify our experience, resolving the grays of reality into the stark black and white of clear recognizability.

An understanding of prototype and caricature is extremely important in the search for the cognitive basis of art, but they are not the only core principles. A number of other artistic rules-of-thumb, such as compositional balance, contrast, repeating forms, and metaphor are also key, but in each case, these are seen to be manifestations of low entropy and/or low algorithmic complexity. Scenes with symmetry and repeating elements, for instance, require less information to produce than asymmetric or non-repeating elements; the latter are comparable to our example of a generator that produces equiprobable symbols, which results in maximum Shannon entropy.

How, specifically, would repeating elements (whether they be in an image, sounds, or idea) be neurally transformed into reward? Consider the mandala image in Fig. 2-9 as an example. The eye fixates on one area of the image, and a visual subassembly is activated. All the feature detectors for the colors, shapes, and their relationships to each other light up

Figure 2-9. Symmetry and repeating patterns, such as in this Tibetan mandala, mean low algorithmic complexity and aesthetic appeal.

with activity. Because attention must shift gaze around the image (the fovea is small), each part of the image needs to be stored in memory while another section is parsed.

To do that, our brains establish a short-term memory trace for the subassembly, wipe the primary sensory cortex clean, and move the eye to inspect another area. Upon reaching it, surprise! The same subassembly is activated. As this process is repeated, the special subassembly will be rhythmically activated over and over at the frequency of attentional flicker, as will the short-term memory trace for it. My argument is that this process excites serotonin-containing neurons, whose activity facilitates dopaminergic reward circuits.

Other artistic principles are more directly related to purely perceptual mechanics, such as stimulus isolation and contrast. High contrast, for example, may indicate that an object is nearby (atmospheric effects reduce contrast with distance) and thus relevant for immediate behavior. Or, a light-dark boundary may indicate edge and shadow, useful for object parsing. Several authors (*e.g.* Enquist & Livingstone) have also pointed out that many biologically important objects, including many other living things, are symmetric and our visual systems might be tuned to look for symmetry for that reason.

Plato thought that painting was at best mimetic: it merely copied the superficial façade of things, utterly failing to make contact with the truth behind the appearance. With most styles of realistic or semi-realistic painting, this holds water, so long as you believe that there *is* any truth behind appearances. As a naturalist, you know how heavily filtered sensory experience is, and therefore what a small subset of nature's deeper truth can be represented by representationalist art. Some art tries to complete a round trip of sorts, first extracting the universals from experience, then producing a particular work that embodies all of them. Different degrees of this distillation and re-expression can be seen in three separate works by Mondrian (Fig. 2-10).

It was in fact exercises like these that resulted in one of the most important developments in the modern history of western art, *i.e.* Cubism, and the realization that one could paint "with the eye," as in the Realism that had predominated western culture for many centuries, or one could paint "with the mind," as we see in the blatantly unrealistic works of, for example, Picasso.

Cubism is the first almost direct attempt to deal with Plato's complaints with mimetic art. It often ignores lighting and fixed perspective, features which, after all, are not

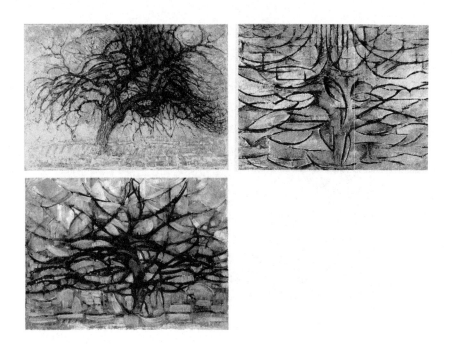

Figure 2-10. Three paintings of trees, all by Mondrian. Painting can be more or less mimetic, but extracting the essence of the tree produces ever greater abstractions.

characteristics of the object in itself, but instead of the artist's limited viewpoint. There is a story in which Picasso was asked by a fellow passenger on a train why he didn't paint people "like they really are." Picasso asked the man what he meant. The man dug from his wallet a photograph of his wife and said "There, you see, that is how she really is." Picasso studied the small piece of paper for a moment and responded, "She's rather small, isn't she? And flat?"[16]

When we consider movements like abstract impressionism or *De Stijl*, it seems that some artists are going even further, going as far as they can past the façade of visual appearances to the fundamental essence of things. Zeki makes

Figure 2-11. Van Doesburg, *Counter-Composition V*, in the De Stijl style.

a case that what they may actually be doing, in part, is getting in touch with the representations of early visual subsystems.

This extraction of essences need not be limited to the visual, but can also include psychological features. Zeki claims that the beauty of Vermeer's painting, for example, is its ambiguity, which allows it to embody a number of simple, almost elemental dispositions. I would argue that the realistic style also lulls us into the belief that we are not looking at a painting but at a real scene. What results is a collection of diminutive micro conflicts and micro resolutions in our attempt to understand the situation. Each resolution is really the recognition of an eternal mood (which Ramachandran identifies as the *rasa* of Hindu art) of some kind, many of which are so subtle that they have no name. The more effective the representation of the artwork, the stronger our recognition of its universality. The binding associated with that recognition gives us, each time it happens, a little bit of pleasure.

By codifying in tangible works the essence of internal representations, art becomes, essentially, a theory of us, of our internal universe. Science, on the other hand, codifies the rules governing natural *external* processes, and becomes a theory of the world. And, as ever, the search for beauty guides us in the development of each.

Aesthetics in science

> If the equations of physics are not mathematically beautiful that denotes an imperfection, and it means that the theory is at fault and needs improvement. There are occasions when mathematical beauty should take priority over agreement with experiment.
>
> – Paul Dirac (1963)

I will provisionally define "theory" as "a simplifying explanation for observations." As explained above, we can think of theories as static and dynamic. Static theories are classification schemes, asserting that some particular animal is a dog. Dynamic theories explain things that change over time, like the motion of an airborne ball. Scientific theories are of both sorts: the Linnean system of biological classification is a static theory, Newtonian mechanics is a dynamical theory. In either case, what makes a theory good?

First, a good theory is *accurate, i.e.* it makes correct predictions. If the animal is a dog, it should bark, not quack. If some dogs can run 100 mph, the theory should include that information. In the dynamic realm, a good theory of ballistic motion should explain why a ball deviates from a parabola, *e.g.* because it has a spin. Second, a good theory is *concise*, explaining a lot with a little. Conversely, a theory may be bad because its predictions are wrong, or because it is just as complex as the original data. The more accurate and concise we can be, the more valuable and appealing the theory.

Other ways to think about these two criteria include the business-like "effective *vs.* efficient" spectrum, the computer-science-like "algorithmic complexity" measure, and of course the Occam's razor perspective. Good theories are effective, efficient, algorithmically simple, and entail a minimum number of assumptions. As an example of a good theory, we'll consider one of the most celebrated scientific developments of all time, celestial mechanics. I include this not as a lesson in the history of science, but rather to show the influence of these two measures of aesthetics in scientific practice.

As far back as anyone knows, humans observed the skies and must have noted certain salient features. The sun's movement across the sky, the phases of the moon, and the differences between planets and fixed stars were all questions begging to be answered. In many cultures, mythologies arose to explain these observations – and these are, despite their supernatural tenets, theories – but eventually, more careful and systematic record-keeping allowed the formation of mechanistic explanations. Some of this was driven by technological need: the invention of agriculture in particular drove the need for accurate seasonal predictions. By the time of Ptolemy (c. 150 CE) the earth-centered geocentric model had been advanced, and although it incorrectly placed the earth at the center of the universe, its predictions were quite good.

The Ptolemaic system was not only appealing for its predictive value; geocentrism made intuitive sense at that time, and was emotionally appealing to those who placed themselves in positions of utmost importance in the cosmos. These attractions stabilized the Ptolemaic theory for an astonishing 1400 years, some of that thanks to the ideological support of the Catholic church. In 1514, however, Copernicus published an improved version of the theory that placed the sun at the center. Though counterintuitive, this model explained the data more accurately and thus probably would have been accepted more quickly if it had not contradicted the still-Ptolemaic Christian version.

In 1609, Kepler refined the theory yet again, providing his famous three laws. These represented advances in both effectiveness (ability to explain the data) and, even more so, in efficiency (compactness of representation). In just three simple equations, everything that had been discovered to date was explained. No need for vast tables of planetary motion year after year…all the positions and apparent motions could be calculated from the three laws.

Finally (for our purposes, though not the true end of the story), Newton formulated his consummate theory of mechanics, which not only subsumed Kepler, and thereby explained planetary motion, but also explained the motion of every massive body in the universe (again, not really, but pretty close).

Through each of these steps, we see increases in theory effectiveness and/or efficiency. Armed with the idea that beauty is the feeling we get when the entropy of our experience decreases, we can also see that the reason each successive theory supplanted its predecessor is that it felt better to think about.

So: theories are tools. They are tools to achieve a certain end, *i.e.* to explain the origin of data, and to make predictions about what will happen in the future. Does this identity work the other way around? Are tools – physical tools – theories? Is it useful to think of a knife as a kind of theory? My answer is yes. A knife is also a tool to achieve a certain end. We can think of a knife as a physically expressed theory of sharpness, just as an artwork is a physical embodiment of a mood and as Newtonian mechanics is a conceptual embodiment of celestial behavior. Thinking of physical tools in this way allows us to place our ideas about the cortical mechanisms of theory generation into the context of human evolution, which is tellingly indexed by the sophistication of the tools they used.

Our temporal cortex is responsible for categorization and thereby inferring hidden attributes. This same skill is present in other animals. Our *frontal* cortex seems to be responsible for predicting things and for generating plans to achieve certain ends. It's therefore a reasonable bet that somehow, the frontal cortex is responsible for generating those other kinds of theories that explain outcomes and predict behaviors, like Newton's laws predict the motion of the planets. And just as the frontal cortex reaches for simplicity as it plans our behavior, eliminating unnecessary movement, so it favors theories that minimize redundancy, inefficiency, and spurious assumptions. In the simplest terms, the biological mandate to save calories has been parlayed into a desire for theoretical parsimony. In other words, the frontal cortex is the neurobiological home of Occam's razor. *Entia non sunt multiplicanda praeter necessitatem.*[17]

In addition to effectiveness and efficiency, a theory can win extra points through many of the same tricks available to other stimuli. Internal structure, like the symmetries found in some of spatial designs we've considered, or the temporal regularities of rhythm and rhyme, can stimulate our beauty circuits and make us like the theory containing those symmetries more. Of course, if the theories are judged by scientists, they must remain strong in the foundational principles of explanatory power and parsimony.

If someone *other* than a scientist is evaluating the theory, however, then explanatory power is not necessarily required. Astrology is an example of a theory with good internal structure, compact premises, a broad scope of application, and an elegant notation that invites us to learn more. This multi-level aesthetic appeal has attracted and held the attention of many people for many centuries.

Unfortunately, the premises are false and the explanatory power is essentially zero.

The classic example on the other side, *i.e.* a theory shining in brevity and accuracy, and also glowing with symmetry and notational elegance, is the four-pillared description of classical electromagnetism which we now call Maxwell's equations. In 1864, James Clerk Maxwell compiled and extended four laws that had been known previously but never unified. These laws and their applications were as follows:

Law	Application
Gauss's law	Electric charges produce electric fields
Gauss's law for magnetism	Asserts that there are no magnetic "charges," *i.e.* monopoles
Faraday's law of induction	Changing magnetic fields produce electric fields
Ampere's law	Electric currents produce magnetic fields

Table 2-1. The compiled laws of classical electromagnetism.

Although there are many mathematical formulations for different physical settings, the appearance of the equations in a vacuum with no currents or charges has a striking minimalism, even to a non-physicist. I present these equations here not only because their omission in this context would be grievous, but also in hopes that they will induce a non-mathematical reader to plumb their depths with further learning. Without further ado (and this order corresponds to the laws listed in the previous table):

$$\nabla \cdot E = 0$$

$$\nabla \cdot B = 0$$

$$\nabla \times E = -\mu_0 \frac{\partial B}{\partial t}$$

$$\nabla \times B = \varepsilon_0 \frac{\partial E}{\partial t}$$

Table 2-2. Maxwell's equations (in a vacuum).

Although Maxwell didn't write the equations in quite this way, the mathematics was the same, and he was able to solve them as a system for the first time. What he noticed was that the reciprocity between electric and magnetic fields produced a self-perpetuating wave whose velocity, which depends only on the two constants appearing in these equations, could be predicted by obtaining values in relatively straightforward experiments with electric currents and magnets. Using such experimentally obtained values, he determined that the predicted electromagnetic wave would propagate through empty space at a speed of 310,740,000 m/s. In Maxwell's words:

> This velocity is so nearly that of light, that it seems we have strong reason to conclude that light itself (including radiant heat, and other radiations if any) is an electromagnetic disturbance in the form of waves propagated through the electromagnetic field according to electromagnetic laws.

There are three staggering observations to make here. First, even if you don't follow the math, it's *visually* clear that the formulation above is highly symmetric. The zeros, the dots, crosses, triangles...they just look *balanced*. If you *do* understand the math, it's even more satisfying, because the symmetries are even deeper than they appear. There are semantics here, as well as syntax, and beauty is hidden in the meaning as well as the form. The second thing to notice is that the laws look not just symmetric, but minimalist. In fact, they are almost *too* simple – as if there just aren't enough symbols to mean anything very important. And finally, as is evident from Maxwell's statement, an enormous range of physical experience can be explained with these laws. Although the laws themselves only explicitly mention electricity and magnetism, the discovery that the wave velocity was equal to the speed of light meant that these equations governed all forms of light, with applications from the Fischer-Price® Easy-Bake Oven (infrared light) to the Sunday paper (visible light) to dentist appointments (x-rays) and beyond. The range of this theory is enormous.

So we must ask ourselves: does such internal structure actually increase the "value" – whatever that means – of the theory? Yes and no. Symmetry, for one thing, suggests that there is actually a tighter formulation that lacks that symmetry. Given *any* data, internal structure tells you, "Wait a minute, these data can be compressed. There is a tighter structure in which the redundant element producing the pattern is eliminated." So if a theory itself has a symmetry (which is really one kind of redundancy), there should be a more compact second-order theory that generates this one. It's that second one that we're really after, because it's more parsimonious. If one of Euclid's axioms were shown to be redundant with another, someone should and would come along to provide a new axiomatization eliminating the redundancy. So, theories that have internal structure are psychologically appealing, but they suggest that an even better one is out there.

The surprise, as many have noted, is that the universe seems to feel the same way. We do not yet know of any obvious reason why an elegant natural law would be more likely to be "true" than a more tangled and incomprehensible one, but so far, elegance in fact seems to have the upper hand. One idea goes at least part of the way in explaining this. The idea is Noether's Theorem, and it says this: there is a one-to-one correspondence between symmetries and conserved quantities, such as:

- Energy is conserved iff (if and only if) the physical laws are invariant with respect to time.
- Momentum is conserved iff the physical laws are invariant with respect to position.
- Angular momentum is conserved iff the physical laws are invariant with respect to rotations.

In physics, conservation laws are as fundamental as they come. For a quantity to be conserved, there must not be any sources or sinks, *i.e.* any places where this thing, whatever it is, can just appear out of nowhere or disappear to nowhere. But where would that "nowhere" be? By definition, it would have to be somewhere outside of nature. If it were in nature, the quantity wouldn't be disappearing "off the books," but would just turn up in another part of the system. Conservation laws might therefore be thought of as corroboration of the physicalist worldview. They are also reassuring in some of the same ways as dimensionless constants, which tell us that the mathematics we use to describe nature is really universal – it doesn't matter whether we measure length in meters or inches, π is always going to be 3.1415...

Seen in this way, Noether's Theorem connects not just conserved quantities and symmetry, but physicalism and symmetry. Basically, Noether's Theorem says that the universe *has* to be beautiful. Of course, one can take the road to simplicity too far. The "God did it" explanation of Intelligent Design, for example, seems a simple answer – certainly simpler at first glance than molecular genetics – but that apparent gain disappears when the terms are analyzed more carefully. As we'll see in Chapter 5, there's nothing simple, or even logically consistent, about invoking an omnipotent deity to explain a natural phenomenon. Moreover, our desire for simple theories must be weighed against their explanatory power. This is algorithmic complexity again. It is always possible to provide a simple program. The trick is to provide a simple program *that generates the data provided* – or, in this case, observed.

"The most incomprehensible thing about the world is that it is comprehensible."

– Albert Einstein

"Simplicity is the final achievement. After one has played a vast quantity of notes and more notes, it is simplicity that emerges as the crowning reward of art."

– Frederic Chopin

In the quest for simplicity, some believe that the physical sciences are today approaching a milestone. The two great theories of modern physics, quantum field theory and general relativity, are known to be incompatible. Until they are reconciled, we cannot really say that we understand the basic structure of the universe. If they are reconciled, a possibility that many find profoundly intellectually and aesthetically satisfying, the result will be a so-called "theory of everything." String theory is a contender in that effort, and it's quite exciting that it may be resolved within the lifetimes of its current practitioners. One of the problems with string theory, however, is that there seems to be an infinite number of different theories that all generate the same universe. This is profoundly unsatisfactory, because the desire for the closure that string theory promises is ultimately a desire for causality. If the universe could have been some other way but isn't, the question naturally arises as to *why* it is not that way. If the answer comes back that *this* universe is but one of many possible universes that could have resulted from the Big Bang and the same set of physical laws and initial conditions, then there is arbitrariness at the foundation of nature. This particular universe would then exist literally for no reason, *i.e.*

it would *have no cause*. That seems to be an escape from logic (and see Appendix B for why such an escape is problematic).

Theists are accustomed to dead ends like this, because God Himself is supposed to be a first or causeless cause (the "Cosmological argument"). Our cognitive machinery is so hard-wired to require causes, however, that the only way to be satisfied with this "argument" is to avoid thinking and simply believe. For a skeptic, such faith is equivalent to giving up on the whole inquiry, and if we were going to do that, what was the purpose of spending so much time and effort getting to this point? And yet, what can we do? Before slipping into despair, keep in mind that it is possible that the need for a first cause may be proven superfluous if the universe we live in is proven to be the only possible, *i.e.* the only logically consistent universe. As a failsafe mindset, however, there may be some value in exploring the origin of our discomfort with the other possibility. Why do we believe in causes? Answering that question will help in our ongoing clarification of the difference between the scientific and religious worldviews.

The prevailing theories of causality – and this is a branch of the philosophy of science, not science *per se* – rest upon the idea of stability. Not mechanical stability, but permanence of identity. If we believe in the stability of a pool ball's trajectory, for example, we are endowing the trajectory with an identity; the direction of the trajectory is what makes it *it*, and we believe that that identity is stable over time. If it proves not to be stable, our belief is violated, and we go looking for a reason. *I.e.* if we observe a pool ball to suddenly change course in its path across a table, we go looking for a warp in the table, a bump to the edge, a hidden magnet, *etc.* If we believe in the stability of a quantity of heat, we believe that the amount of heat is what makes it *it*, and any change in that identity encourages us to look for a reason; things don't get warmer or cooler without a cause. Thus the conservation laws found at the fulcrum of Noether's theorem, which in turn form a basis for all known physical law, are intimately related to the philosophical notion of causality.

What this underlying belief in stability amounts to, however, is a belief in simplicity. The *alternative* to believing that the pool ball's trajectory has a stable identity is to believe that it does not, but to do that is to believe that the trajectory can change from moment to moment, perhaps capriciously. The ball points this way, then that, then back again. Even if this happened so rapidly that we could not measure the variations, and even if all those changes eventually cancelled out to make the trajectory simply *appear* to be stable (as in Feynman's sum over histories approach to quantum mechanics, for instance), this would still require the provision of information in the Shannon sense, or a more complex algorithm in the KC sense. A single value for the trajectory is the most parsimonious, least entropic state of affairs. Any additional values add complexity. Thus, our belief in causality ultimately reduces to the assumption-minimizing dictum of Occam's razor, a.k.a. our emotional preference for low-entropy stimuli, which is the result of serotonin-mediated reward when strong gamma rhythms are established quickly. Once again, we return to evolutionarily-motivated neurobiology and find a candidate mechanism for how its microscopic dynamics give rise to some of our most sophisticated endeavors. The practice of science is not a metaphysical quest at its core. It is a behavior with idiosyncrasies rooted in biology. Nevertheless, Einstein was right – the really wondrous thing is that, despite its gooey origins, science does seem to be bringing our mental models into register with the world we observe.

The religious experience

We have seen how the evolutionary quest for order is realized through a diverse range of human activities. Science and art are unified as efforts to decrease the entropy of experience, and as conscious strivings toward the aesthetically pleasurable. Poetry and music do the same and rely even more explicitly on the rhythmic phenomena that lie at the heart of the neural hypothesis I have proposed. Religion too is a quest for order, albeit one that is methodologically somewhere between science and art: it claims truth but does not subject itself to logical or empirical constraints. With the theoretic products of science lying like waypoints on the path to a theory of everything, we might draw the parallel to religion and ask, "Is there a neural and evolutionary framework that explains the goals of religion as well?" If we can answer that question, we will have gone a long way toward addressing one of the stated goals of this book.

The western analytical study of religious and mystical states began with William James' 1902 publication of *The Varieties of Religious Experience*. As a psychologist and philosopher even before psychology became known as a field, he was a pioneer in being the first to isolate and carefully describe first-hand accounts of these unusual states of mind. A representative quote from one of the anecdotes he collected from subjects reads as follows:

> "In that time the consciousness of God's nearness came to me sometimes. I say God, to describe what is indescribable. A presence, I might say, yet that is too suggestive of personality, and the moments of which I speak did not hold the consciousness of a personality, but something in myself made me feel myself a part of something bigger than I, that was controlling. I felt myself one with the grass, the trees, birds, insects, everything in Nature. I exulted in the mere fact of existence, of being a part of it all – the drizzling rain, the shadows of the clouds, the tree-trunks, and so on." (p. 394)

Other subjects describe "the vanishing of the sense of self, and the feeling of immediate unity with the object," the "innate feeling that everything I see has a meaning," and the sense that "Knowledge and Love are One." James summarizes:

> This overcoming of all the usual barriers between the individual and the Absolute is the great mystic achievement. In mystic states we both become one with the Absolute and we become aware of our oneness. This is the everlasting and triumphant mystical tradition, hardly altered by differences of clime or creed. In Hinduism, in Neoplatonism, in Sufism, in Christian mysticism, in Whitmanism, we find the same recurring note, so that there is about mystical utterances an eternal unanimity which ought to make a critic stop and think, and which brings it about that the mystical classics have, as has been said, neither birthday nor native land. Perpetually telling of the unity of man with God, their speech antedates languages, and they do not grow old. (p. 419)

Many years later, in 1975, Stephen Waxman and Norman Geschwind, neurologists at Yale University, furthered the field when they published a report about a group of patients with temporal lobe epilepsy (TLE) who experienced frequent experiences like those described by James, and became obsessed by them. They became hyperreligious. They also had other symptoms, such as hypergraphia, the tendency to write prolific amounts. Since then, yet more traits have been added to what is sometimes called the Geschwind Syndrome or the *interictal personality*. For our purposes, I will mention just two of the diagnostic symptoms: (1) obsessive interest with deep philosophical speculation or theories of cosmological significance; and (2) a tendency to ritual and repetition. Patients with these traits are still found. Here is the neurologist Ramachandran quoting and describing one of his TLE patients in 1999:

"I had my first seizure when I was eight years old," he began. "I remember seeing a bright light before I fell on the ground and wondering where it came from." A few years later, he had several additional seizures that transformed his whole life. "Suddenly, it was all crystal clear to me, doctor," he explained. "There was no longer any doubt anymore." He experienced a rapture beside which everything else paled. In the rapture was a clarity, an apprehension of the divine – no categories, no boundaries, just a Oneness with the Creator. ...

The next day Paul returned to my office carrying an enormous manuscript bound in an ornate green dust jacket – a project he had been working on for several months. It set out his views on philosophy, mysticism and religion; the nature of the trinity; the iconography of the Star of David; elaborate drawings depicting spiritual themes, strange mystical symbols and maps.[18]

Particularly due to their connection with religious experience, such accounts have generated a good deal of controversy, and probably did so long before 1902. Even so, there is considerable disagreement about TLE's causative role in this cluster of symptoms, or even whether there is such a cluster. The controversy gets additional steam from the long-known association between religious states and serotonergic hallucinogens, although (strangely, perhaps) this latter link is more thoroughly documented.[19] These are all irresistible topics for popular audiences, but I can assure you that talking about mysticism and LSD in the hallowed halls of academia is a very efficient way of evading tenure. This is probably one reason why the scholarly literature is a bit thin.

On the other hand, anyone who has actually experienced such a state will attest to its overpowering emotional and even intellectual significance. We willingly subject far more mundane mental experiences to scientific study; why does one that we find so meaningful not deserve equal attention? I direct this question both to scientists and general audiences, because scientists are the ones reluctant to study it, but the general public's influence on science is meaningful. In any case, one might guess that there are several reasons for the relative paucity of work to date. These include (a) the expectation that the mystical experience must be extremely complex neurologically and therefore too hard to study; (b) the societal norm that religion is a private matter, inappropriate for open discussion; (c) the feeling that, because religion is connected with faith, such experiences do not submit to rational inquiry; and (d) fear of angering religious extremists. Some of this is akin to what Kinsey faced with his study of human sexuality, some of it is a conflation of religion with the religious experience, and the rest is just lack of nerve. As James intrepidly states,

> I do not see why a critical Science of Religions of this sort might not eventually command as general a public adhesion as is commanded by a physical science. Even the personally non-religious might accept its conclusions on trust, much as blind persons now accept the facts of optics--it might appear as foolish to refuse them. Yet as the science of optics has to be fed in the first instance, and continually verified later, by facts experienced by seeing persons; so the science of religions would depend for its original material on facts of personal experience, and would have to square itself with personal experience through all its critical reconstructions. (p.456)

Sharing that conviction, we can now speculate on what our theory of beauty might have to contribute to this strange field.

Binding individual components to make a unified percept is essential for normal sensation, *e.g.* to see an apple as an apple instead of a collage of spatially contiguous but functionless parts. You can think of this as a top-down force favoring unitary percepts over exploded collections of features. This same process happens during the formation of categories. There is a pressure to see an individual apple as just one member of the *apple* category. This is enormously powerful since, as I have said, it allows you to make inferences about hidden attributes.

In turn, I have explained a possible mechanism for how reward is produced when the individual apples – or whatever – are close to the category prototype. I also discussed how language itself may use prototypes as its basis. When I say "apple," you picture a generic apple, not the particular one you had for lunch last May. (This appears not to be true, however, for autistics. More on that in a minute).

One might also productively think of dynamic theories as cognitive tools that unify chunks of experience under one predictive idea, just like categories. Newtonian theory unifies celestial mechanics with ballistic motion, for example, and as such is explanatorily powerful, algorithmically simple, and aesthetically appealing. Same for Maxwell. All of these processes are characterized by taking a diverse body of experience and condensing it into a simpler, more unitary scheme. The link between this process and reward, or pleasure, is a fundamental piece of the theory I am proposing.

It's worth mentioning, too, that the symmetry of a mandala, the commutativity of elegant equations, the rhythm of a drum and the rhyme of a poem also produce pleasure because each has a simple core that explains or generates something more complex. There is *one* template in the kaleidoscope, *one* symmetry in Maxwell, *one* pattern, *one* scheme. In other words, we unify features into objects, objects into categories, experiences into predictive theories. It is the simplifying, entropy-reducing process of unification that provides cognitive advantage, evolutionary value, and the positive emotion we feel when we're successful.

But far from being some hidden feature of aesthetics, our desire for unification is quite explicit. In science, "Theories of Everything." In religion, the Holy Trinity or Brahman = Atman. In politics and popular culture, one world, one vision, one truth, one voice, John Lennon's "Come together," the United Way, and on and on. The idea of bringing everything or everyone together into a single construct is immensely appealing, even mystical. We say things like "It all just came together for me," or "It all comes from the same place," or "It all comes down to this." And these statements are not associated with disgust or frustration or hunger – though in principle they must just as well have been – instead they are associated with feelings of closure and satisfaction.

> Three Rings for the Elven-kings under the sky,
> Seven for the Dwarf-lords in their halls of stone,
> Nine for Mortal Men doomed to die,
> One for the Dark Lord on his dark throne
> In the Land of Mordor where the Shadows lie.
> One Ring to rule them all, One Ring to find them,
> One Ring to bring them all and in the darkness bind them
> In the Land of Mordor where the Shadows lie.
>
> – J. R. R. Tolkien, *The Lord of the Rings*

Simplicity. Unity. Beauty. Reward. Hold that thought.

Above, I summarized the temporal correlation hypothesis, which says that the binding together of disparate components into unified percepts might be accomplished via the formation of dynamic circuits of neurons firing rhythmically and simultaneously. In other words, when the brain needs to perceptually or cognitively unify features or ideas represented in disparate areas, it sets up rhythmically firing sets of cells, also called "cell assemblies" or "neural ensembles."

With seizure disorders, including temporal lobe epilepsy, the problem is that sets of cells start firing together rhythmically, but without the stimuli that would be required for a normal brain. Also the number of neurons included in the synchronized set spreads, recruiting more and more into the ensemble until, in the most pronounced cases, *i.e. grand mal* seizures, nearly the whole brain fires together in rhythmic bursts of action potentials. *Grand mal* seizures are relatively rare; more common would be a seizure locus occupying just a small area of neural tissue. In the case of TLE, the focus is somewhere in the temporal lobe (which, incidentally, is where many memories are stored and the semantic aspects of language are generated and processed).

Now we come to the synthesis. Small sets of oscillating cells are used to represent the unification of features into objects, and this is how we perceive normal stimuli, like apples. And just as the apple itself needs to be glued together, it needs to be identified with the memory of an apple. Memories are stored in the connections between neurons, and if a stimulus induces activity that is supported by those synaptic weights, then strong gamma coherence will be achieved rapidly. Thus the strength of gamma coherence and the rapidity of its establishment can be used as a measure of familiarity, and by prior arguments, it can also be used to judge prototypicality and aesthetic appeal. But if gamma coherence is fundamentally the metric of interest, then it doesn't matter how it is induced. Whether it is induced by perception and recognition of something familiar or low-entropy, or instead by a pathological state, *e.g.* a seizure, or by entrainment to an external periodic stimulus, is immaterial. The subjective consequences will be the same in either case.

Given that, what kind of thing would you perceive if, instead of just a small set of feature detectors, large swaths of your cortex began to fire rhythmically, in one coherent neural ensemble? Not just the sensory feature detectors for a few discrete line segments or colors, but the entirety of atomic sensory and cognitive representations. If these were all bound together into a single neural ensemble, what might you say about the would-be corresponding world-object? And second, given the unusually large amplitude of the oscillation, and the ostensibly high serotonergic activations, what would be your subjective interpretation of the familiarity and beauty of the object? How would it make you feel?

First, it would be quite a chimera. All colors, all textures, all voices, all smells, the whole of sensory experience would be wrapped up, Willy Wonka style, into one weirdly multifaceted gumball. Most everyday perceptions are relatively small collections of bound features. The more complex the object, the more features are brought into the binding circuit, and this object would have a huge set of features, perhaps every feature, because so many neurons would be part of the bound circuit. Every actual object that you could potentially see in the world would be a subset of this all-encompassing thing. You might, in the grip of this experience, have the impression that ordinary objects *strive to be* this thing but do not quite achieve it. And yet, pieces of it are scattered like shards, hidden in the visible world. Is this mega-object familiar to you? You have caught glimpses of it your entire life, but never seen it unobscured, in the light; it is the completion of all experience.[20]

If you knew the particulars of a given nervous system and had sufficient resources, it would be possible, given a neural signal, to decode what stimulus produced it. Recall from our information theory discussion that this has been done in several invertebrates with great success. Using slightly different techniques, the same thing is nearly possible in primates and measurement of the activity of their primary visual cortex. That is, it is *almost* possible to construct a line drawing of the original scene, given just the firing rates of cells in visual cortex. As a thought experiment, suppose we were able to do this projection from neural activation space to perceptual space for a random set of bound elements. Most likely, the corresponding perceptual object would in general be physically unrealizable, or at least unnatural. This is because the world is so constrained. As Dali said, "so little of what might happen does happen." In other words, imagine picking ten feature detectors at random and forcing them to fire simultaneously. What world-object could produce just this pattern of activity? It would be quite surprising if it were some object that actually appears in the world on a regular basis; its edges would probably not line up the way they should for a real world object, colors would probably be discontinuous, sounds would come from areas distant from their supposed source...more likely than not, the reconstructed stimulus would be a bizarre impossibility rather than a mundane everyday thing. So, yes: if, in this hypothetical seizure state, all feature detectors, not just visual, but all senses, became bound together into one percept, it would be rather extraordinary.

And what emotion would this grand percept / concept produce? Reasoning by inference from the way more ordinary perception is handled, we could reasonably expect that a global feature binding of this type might produce euphoria. Why? In small doses, strong establishment of gamma oscillations signify recognition, and this has been made rewarding by evolution. When gamma oscillations are strong, serotonergic activation and reward are high *irrespective of how those oscillations are established*. In the case of a seizure, an exceptionally large swath of cortex is involved in one coherent binding circuit, which might be interpreted as truly magnificent categorical recognition and / or theoretic insight. Limited only by the size of the epileptic focus, the whole of experience would seem familiar, fitting neatly into one infinitely complex (because it is so multi-featured) and yet infinitely simple (an interpretation based on the strength of the gamma rhythm) nameless thing. – Or perhaps it is not so nameless? Brahma. YHWH. Allah. God.

> But when that which is perfect is come,
> then that which is in part shall be done away.
> When I was a child, I spake as a child,
> I understood as a child, I thought as a child:

but when I became a man, I put away
childish things. For now we see through a
glass, darkly; but then face to face: now
I know in part; but then shall I know even
as also I am known.
                    – 1 Corinthians 13:10-12

Simplicity. Unity. Beauty. Rhythm. Reward. God.

If the construction that has brought us to this point seems unstable, consider this. The keystones in this argument are serotonin and neural rhythmicity. Serotonin was brought in because it seemed well-positioned to be the link between sensory binding, rhythmicity, and reward. The case for serotonin and sensory binding is fairly strong because we know that most hallucinogens act uniquely at a single type of serotonin receptor, 5-HT_2. The case for serotonin and reward is also pretty well established due to the anti-depressive effects of serotonin-facilitating drugs. The serotonin-to-rhythmicity case is admittedly not quite as strong, but the cluster of data surrounding LSD, rhythm, neurological disorders, antidepressants, *etc.* makes a good circumstantial case. Wouldn't it strengthen the argument further if there were data which skipped the entire construction and showed a direct correlation between serotonin and religious experience itself? It would. In 2004, NIH geneticist Dean Hamer analyzed the DNA of people ranking highly on tests of spirituality and self-transcendence. What he found was a correlation between high scores on that test and the possession of a unique form of a gene called VMAT2, which stands for "vesicular monoamine transporter 2." Basically, the gene makes a protein responsible for regulating the amount of certain types of neurotransmitters – the monoamines – stored in cells. And is serotonin a monoamine? You bet. At this point, not much more is known, but it is supportive of our hypothesis that, of all the thousands of proteins that VMAT2 might have coded for, it just happens to code for a chemical regulating serotonin.

*   *   *

At this point in our investigation, we've assembled a body of data and theory suggesting a purely naturalistic, *i.e.* physical and analytical basis for the canonical unification experience and for the desire to understand the world simply. When such states manifest very strongly, the experience might be interpreted as mystical or religious, or, in more moderate settings, as aesthetic or artistic. Before closing this section of the book, it's worth touching base with two final aspects of this quest. First is a religious / mystical experience that is not based on unification, namely the nihilistic one of Eastern traditions, and second is the tendency toward symbolism *per se,* not just in religion but also in its storytelling and saga-spinning precursor of mythology. The first of these considerations will close this section, and the second will close the chapter.

Due to its foundation in the gamma-bound ensembles encompassing large swaths of cortex, another way of thinking about God is that it is the *union* of all sensory experience. This leads naturally to a second question: What are the characteristics of the percept or world-object corresponding to the *intersection* of all sensory experience? Might that also be a form of religious experience?

To consider *this* object, imagine collecting all sensory experience over a lifetime and asking whether there is some set of neural activity present in every case, some universally shared quality. And then, what does that "thing" look like? What does it seem like to perceive it? Arguing by analogy, I claim that such an object (well, not really an object) does exist, and that its name is *Is*. Very tidy! The one quality that all perceivable objects share is that they *exist*. This is like the epistemological extension of the Cartesian proof of existence of self. If you perceive something, then that something "exists," if only in your mind. This *Is* object is the other end of experience, the second boundary condition, and with God, it brackets all the rest of perception.

We might also say, from a KC complexity point of view, that God is the algorithm that generates the world, a.k.a.

a theory of everything. This would seem to coincide with the God of Spinoza and Einstein, the God "who reveals himself in the orderly harmonies of the universe." The other end of the spectrum is of course a program that does nothing: {}. Instantiation of this program is the nihilistic experience, in which categorization is turned off, theorizing is turned off, language is suppressed, the mind is quieted and the quest for order ceased. This is nirvana.

Symbols and myths

The tendency to ask, "What does x mean?" where x is life, a dream, an event, *etc.* is simply an overextension of our linguistic and symbolic instincts. It translates to "What is x a sign of?" or "What is signified by x?" This is a fallacy: not everything is a symbol. Or more accurately, *nothing* in the world is a symbol. For nouns like "apple," "chair," and "brick," the associated object produces a representation in your brain by exciting some set of peripheral sensory neurons. Other nouns designate things that never existed as objects in the peripheral sensory world in the first place. For example, "guilt," "sorrow," and "dedication." You might be able to infer from someone's behavior that they are experiencing one of these things, but there is not a thing out there that you can point to. Instead, these words denote subtle principles extracted from your own emotional responses to the world.

Does this mean that we are limited to using words as symbols? No. We can use objects to represent other objects. For example, I might lay out a military strategy using stones and twigs, saying that this pebble is the 4th infantry, this stick a Panzer division, *etc.* Almost anything can be used as a symbol. Some things are of course more natural to use than others. It would be possible but unusual to say that, in some exercise, the concept of guilt stood for a mushroom pizza. One the other hand, the reverse, a mushroom pizza standing for guilt, seems a little eccentric but not deeply alien. It could be that natural-seeming symbolic designations have the shared property that they denote complex objects with simpler ones (information loss), not simple objects with complex ones (information gain).

Suppose I say that a small rock stands for happiness. Any time I use an actual object, *i.e.* something in the peripheral sensory world, as a symbol, that symbol becomes an *icon*. By using icons, I can create allegorical tales that superficially are about sets of fairly simple objects, but when decoded and mapped onto their respective referents, are seen to be more complicated communications about values, morals, or whatever. This is a favorite human pastime, resulting in the collection of fables, myths, and legends of cultures going back as far as we can see.

Aside from using these devices consciously to entertain and educate, we may use them unconsciously and perhaps even unwillingly, distastefully, in constructing views and opinions of events and states of affairs. For example, a person, who is after all just a person, can come to symbolize a virtue, vice, or force. There are many examples. Most apropos to modern life is the phenomenon of media and entertainment celebrities, particularly actors and musicians, who seem to be embodied caricatures of some admirable trait. If the person is not actually a human being but rather a humanoid character in drama, we obtain deities. There is Mars, the Roman god of war; Pele, the Hawaiian goddess of fire; Shiva, the Hindu god of destruction; Osiris, Egyptian god of the dead. Each of these tends to be an incarnation of some set of values or morals. We can use stories about them to inculcate children into the norms of social morality and worldview and also to help make sense of unpredictable world events ourselves. Why is that volcano erupting? Because we have not made sufficiently many sacrifices at Pele's temple.

What is particularly interesting is when icons are really just everyday mortals. When such people facilitate their own iconization, an admiring public can turn them into figures of superhuman importance and stature. Christ and Buddha are two available examples.[21]

Another aspect of religion that can facilitate mystical experience is ritual. This has historically been a universal feature, embraced by belief systems as diverse as Orthodox Catholicism, Islam, Judaism, and the tribal religions of Papua New Guinea and Amazonia. As with community development, moral education, and music, it is one of those aspects of religion that has nothing to do with faith. In fact, ritual seems to be rewarding in its own right, and probably pre-dates all forms of religion, having been co-opted by it as a useful tool. Secular holiday customs, reading the Sunday paper, and the canons of a high school pep rally can all be simultaneously comforting and irreligious. Why?

Recall from Chapter 1 that one of the frequently shared symptoms of autism, OCD, and GTS is a strong need for routine. Deviation from routine is highly aversive, especially for autistics, to the point that furniture must be precisely arranged and the activities of the day performed at precisely the "correct" time. Presumably, there is something about the neurology of these disorders that accounts for this psychological pressure; the desire for similar types of ritual in non-autistics – absent the debilitating anxiety – is probably explicable via the same mechanisms. Although we do not completely understand the neurology of those disorders, we have enough of an idea to devise treatments that are at least somewhat helpful. Those treatments led to the hypothesis already presented above, involving predictability, entropy, serotonin, and reward. What this suggests is that the pleasure most of us derive from ritual is the same as that for autistics – it's just that they require the ritual to feel normal, while for non-autistics, the reward yields hedonic feelings above and beyond normalcy. People with autism and OCD are *obliged* to ritual; for everyone else it is an option.

Of course, ritual is not just about stereotyped behavior, it is about the *meaning* of those behaviors. There is a semantic, not just a syntactic structure. As such, the emotional

Figure 2-12. Ceremonial Jewish Seder plate.

connotations of the ritual's meaning bring to bear the same opportunities for aesthetic appreciation as are available in any other semantic structure, such as prose or poetry.

It might be worth alluding here to another approach to the question of ritual, namely that of existential philosophy, particularly Jean-Paul Sartre. That approach will be discussed at length in Chapter 4, but for now simply note that Sartre's account turns on the idea of freedom and the anxiety produced by forced choice. In other words, ritual absolves us of the need to decide what to do next. The anxiety of not knowing what to do next is exaggerated in the aforementioned disorders, but I will claim that it is, at its core, the same evolutionary pressure for predictability that we have already discussed.

One final note for the appeal of ritual. The *enactment* part of ritual is not the only one of its features whose hedonic value is supported by the hypothesized link between rhythmic neural activity and serotonin-mediated reward. Although ritual is by its nature recurring and therefore macroscopically periodic, there are often smaller-scale regularities nested within it, and these too contribute to the pleasurable associations. The Seder plate pictured in Fig. 2-12 is not only a prop in an annual tradition, but it is in its own right a low-entropy stimulus, appealing for the same reasons as any similar work of art, like one of the grids from above. Music too is often brought to bear, and in tribal cases it is generally very rhythmic. The Yanomamö, the Hopi, and others famously use serotonin-mimicking hallucinogens in their rituals.

I have deliberately focused on the greatest extension of the unification and nihilistic experiences because they are so subjectively different from everyday perception and so central to the mystical aspect of Western and Eastern religious mindsets. As we saw above, the centrality of these spiritual poles is supported by over a century of psychological accounts. Of course, over the same span, there have been countless other, more subtle, more personal experiences such as conversion stories, emotional connections with Jesus and Brahma and Vishnu and other supernatural religious figures, claims of miracles or paranormal events *etc.* These are, however, even less systematically examined, more diverse, and therefore less admitting to scientific investigation, even one as speculative as this. In such cases, I would only care to apply the present hypothesis to the extent that the self-reported phenomenology was similar to that described here.

Other authors have proposed alternative neural theories of religious experience. Andrew Newberg and his colleagues are also interested in the unification experience, but have suggested that its neural origin is in the suppression of activity in the "orientation association area," or OAA, that part of the parietal lobe cortex that processes spatial information and distinguishes the literal physical edge of the self with respect to the rest of the world. Interestingly, they claim that rhythmic activity can interrupt the functioning of the OAA and cause the unification phenomenology. This part of their suggestion, at least, seems to offer a complementary perspective to that presented here. On the other hand, there are significant differences between the proposals; Newberg explicitly discounts the involvement of seizure-like neural activity and hallucinogens in the generation of the unification experience, for example, and believes that its emotional power comes from a connection to the neurological structures involved in orgasm.[22] Part of the reason for this different explanation may come from Newberg's use of brain imaging as an experimental platform. The highest resolution of the SPECT scanners they use is about 1-2 mm, which includes the activity of at least thousands of neurons. Such imaging is extremely useful for many purposes, but it does limit the discussion to a level more macroscopic than the neuron-by-neuron stratum discussed here. Overall, the fact that similar inclinations toward the importance of rhythmic neural activity in religious experience could result from such different approaches is encouraging. As measurement techniques improve, it may become possible to study the intersection of electrical   and   chemical   (*e.g.*   serotonin-mediated) neurodynamics as they relate to religious and even aesthetic experience.

In addition to this neural slant, there are other ways of understanding the origins of religious ideation. Evolutionary and cognitive approaches in particular can be highly productive. Dennett (2006), for example, builds a compelling case and explains a great diversity of phenomena, almost non-overlapping with those selected here. Some opposed to the premise of either this book or Dennett's might argue that they do not adequately address the inherent phenomenology, the "what it is like" aspect of religious experience. Dennett quotes religious scholar Mircea Eliade:

> A religious phenomenon will only be recognized as
> such if it is grasped at its own level, that is to say, if
> it is studied *as* something religious. To try to grasp
> the essence of such a phenomenon by means of
> physiology, psychology, sociology, economics,
> linguistics, art, or any other study is false; it misses
> the one unique and irreducible element in it – the
> element of the sacred.[23]

Contra Eliade, I would agree with both Dennett and Richard
Feynman that far from omitting anything, reductionist and
interdisciplinary explanations of these phenomena only add to
our understanding of the experience of the sacred.

CHAPTER 3

Souls or Neurons?

We have now completed an exposition of the most focused and detailed neurobiological hypothesis of the book. More heavy lifting will follow, but along somewhat different lines. For the moment, we zoom out to consider some of the more obvious questions arising when physicalist neuroscience and supernatural worldviews are juxtaposed. Such questions arise almost automatically when the implications of the last two chapters are fully digested.

Consider for a moment the unlikely possibility that you began reading this book as a fence-straddling theist, but have been so persuaded by rhetorical charm and scientific rigor that you are now a non-theist. You now believe that the God you once saw as a personal and transcendent figure is actually a special state of brain excitation. Even if this has happened, you may still believe in some sort of metaphysical autonomy for yourself. In other words, you may have transferred some of the magic (*i.e.* the something special that exists outside the heartless mechanisms of physics) from God to what you continue to see as the (probably immaterial) human spirit. The most popular way that this transfer of magic is done is to believe, as I said, in transcendent autonomy, or free will.

There is a colloquial meaning of the term "free will," which is this. If you are coerced or forced to do something under threat, you are not doing so of your own free will. Only when you choose a course of action freely are you exercising freedom. This is the street meaning of the term, but it is not what I mean. What I mean by "free will" is more along the lines of how philosophers use the term, *e.g.* "the ability to direct the course of one's own thoughts and actions." In this definition, to believe in free will is to support the idea that your mind is not pushed around by the mere collisions of atoms, the reactions of chemicals, the firings of neurons, but is instead driven by pure *will*. Your mind is a free agent, able to go wherever it wants, whenever it wants. In the relationship between mind and brain, the *mind* is the king, and the brain is the subject.

That's the strong version, sometimes called *libertarian* or *contra-causal* free will. Many people will make concessions against the strong version, trying to acknowledge at least some science, but still stick with it in the very end. Everyone knows, after all, that the brain has some important influence over self, over thought. People in car accidents who suffer brain damage might become partially paralyzed or even have trouble speaking or caring for themselves; the same is true for soldiers with head wounds or grandparents who have had strokes. Memory can be impaired, as can all sorts of higher mental functions, including personality itself. As Heraclitus said, "A blow to the head will confuse a man's thinking, a blow to the foot has no such effect; this cannot be the result of an immaterial soul." So informally at least, many people accept that the strong form of the free will position needs to be softened...the question is just how many concessions to physicalism are reasonable. The mind is *mostly* the boss, they would say, but in special circumstances like brain damage, it might be obstructed in its normal functioning. Or maybe it has a rarely-exercised veto power (amusingly called "free won't") to overcome base biological urges in favor of a moral or altruistic choice.

This is about the point where most people stop.

But this is an inquiry, so let's examine the other extreme: determinism. By this, I mean the idea that one's thoughts and actions are passive consequences of physics. This position is also called *epiphenomenalism* and *causal determinism*. In this view, your mind is just along for the ride. The universe, made up of atoms or quarks or strings, is like a giant computer, with every cubic femtometer of space-time constantly churning through some huge set of calculations embedded in superstrings or whatever the ultimate basis of reality is. Those calculations manifest themselves as chemicals, cells, organs, and organisms, and your brain – one organ inside one of those organisms – is just as mechanistic as a billiard table. Whatever the brain does is entirely a result of the underlying physics, and one of the things the brain does is produce the mind *in its entirety*. As an "epiphenomenon," there is no way for the mind to influence the brain: the causal arrow points from brain to mind but not the other way. Allowing the mind freedom would be like saying that the billiard ball has a choice over which direction it rolls after it is hit.

Those are the two extremes. Many people – scientists, philosophers, and theologians – think both of these positions are partially correct, and are trying to find a way to keep free will on the books but acknowledge some of the aforementioned relationships between brain and mind. That exact debate may be over 2000 years old. Such attempts at reconciliation are misguided, however, and my goal in the rest of this section is to try to convince you that the second extreme, determinism, is the correct view. Many people have strong intuitive feelings that determinism cannot be right. If you're one of them, remember two things. First, intuition is a polite word for unsupported bias. Keep it on a leash. Second, we mustn't base our determination of truth or falsehood on the desirability of a claim's consequences. No matter how emotionally objectionable you find a position, no matter how uncomfortable it makes you, the content of the case must be considered as objectively as possible – to do otherwise would be to commit an "appeal to consequences" fallacy, which we will discuss in more detail in Chapter 5. Later, we actually will consider those emotional consequences, but only in order to put them in context, not to judge the claim on which they are based.

Evidence against free will

The 18th century Scottish philosopher David Hume was the first to point out that even without doing any neuroscience or physics, there seems to be a profound *logical* problem with the very concept of freedom. Suppose that one inclement morning you are deciding whether to go to work or stay home. Presumably that decision will be based on a number of factors other than the weather: your sense of responsibility, the urgency of work-related tasks, the possibility of getting fired if you don't go, and so on. All these factors, some of them innate and some based on experience, will play into the making of your decision.

As you deliberate, you entertain some sort of conscious process in which the pros and cons are weighed against each other. If you are behaving rationally, you will presumably select the course of action that makes the most sense, *i.e.* has the best benefit-to-cost ratio. Another way of saying this, however, is that if someone else knew your character and your experience well enough, they would be able to predict what you would do, because they would also be able to calculate how *you* would calculate the payoffs for each possibility. Knowing the extent to which you weigh rational factors against emotional conditions (which they would also know about), they would know that you would select the path that maximized the overall, *i.e.* rational plus emotional, gain. Since all you would be doing is maximizing a quantity, what is happening is not a choice at all, but a calculation whose outcome could be known if all factors were taken into account. This third-party predictability of what seems to be a free choice is familiar to anyone who has a close friend or spouse. The more they know you, they more they can predict what you're going to do.

The libertarian's intuition here, of course, is that while you might feel *disposed* to go a certain way in such a scenario, you are not *obliged* to do so – this is the veto version of free will mentioned earlier. If you believe in free will, you believe that the rational process outlined above merely sets the stage for the decision – it is up to you, ultimately, to step across the threshold, and you can always choose not to. To investigate that claim, we need to move past Hume into some more modern experimental results.

The first of these comes from Benjamin Libet in 1983, and the experiment is as follows. Subjects are given an EEG cap to wear; this is a flexible mesh that fits over your head like a baseball hat, embedded in which are between 20 and 100 electrodes. These are surface electrodes only, resting against the surface of your skin and picking up whatever electrical activity happens to diffuse out from the neurons inside the skull. Libet had his subjects look at a rotating stopwatch, and instructed them to move their hand whenever they felt the urge. They were additionally instructed to pay attention to the moment, as measured by the position of the rotating watch hand, at which they first felt the urge to move. With these instructions, Libet found that most people reported the first urge to move about 200 msec before they actually began to lift their hand. This seems to comport fairly well with most of our expectations: we get the idea to move, and a short time later, sufficient for the signal to work its way down from the brain, the muscles actually contract. So far, so good.

The problem is that by monitoring the EEG activity on the subjects throughout this process, Libet found clear evidence that the brain was beginning the process of movement about 1250 msec before it actually occurred. That is, a full second before the person's intention to move, some part of the brain had already decided that it was going to do so, and had begun the requisite preparatory neural activity. If something other than intention began the movement-initiation process, then surely intention was *not* the original cause of the movement. The timeline just doesn't work. Instead, the preparatory activity beginning 1250 msec before the movement is the real original cause, and consciousness, in the form of awareness of an intention, only finds out about the plan later on. Just like the muscles of the finger, the conscious urge to move is a passive observer to brain activity.

Libet's result has been replicated by other labs, some of which have extended the finding by using more sophisticated measuring instruments, including fMRI. This has shed light on which specific *parts* of the brain are responsible for that 1250 msec preparatory activity, and has shown which other parts subsequently become active, and when. The main finding here is that a portion of the frontal lobe called the pre-supplementary motor area (pre-SMA) and an area of the parietal lobe both are planning to move before a person is aware of their intention to do so. In fact, patients with lesions to the parietal lobe are even more in the dark about their own intentions – they can still initiate "voluntary" movements, but they only become aware of their intention to do so immediately prior to the movement.

Tightening the vise a little further, several labs have succeeded in *producing* a sensation of conscious will by stimulating these very areas. If free will really existed, it would be causally primary, immune to manipulations of the brain. But this is not the case: one can stimulate the brain and actually make people *want* to do something. They don't feel coerced. They actually *want* to do it, without awareness of any influence or cause. Thus, we know for sure that conscious will can be manipulated by stimulating the brain. Unfortunately for the free-will-ists, there is no evidence that the reverse is true: no one has ever observed brain activity that inexplicably arises from some "higher level" free process.

These would seem to be fairly tight cuffs from which to escape. The most parsimonious explanation is that our intuition about conscious will is wrong. Rather than being in charge of ourselves, we are actually passive observers to a series of calculations done by our brains. To be sure, the outcome of those calculations is determined by our characters and our experiences, and they reflect the influences of years of learning, conditioning, knowledge of what works and what doesn't, and meditative reflection. *But those things too are mechanical processes.* There simply is no evidence for a point at which the physical substrate of brain processes opens itself up to the influence of an immaterial soul or will.

Evidence notwithstanding, many people continue to try to wriggle out of these cuffs in an attempt to find a way to be free. This is perhaps not surprising, because counterintuitive ideas are always hard to accept, even when the evidence is strong. If our conscious will is the result of mechanical processes, such people may say, then surely it is the result of a mechanical process that somehow breaks the shackles of billiard-ball physics and becomes more than the sum of its parts. "I know what it feels like to *have* to do something, and I certainly didn't *have* to have waffles this morning. I could have had eggs or anything else. No way is my mind just blown around by physics. There's no way anyone could predict what I was going to do next...even if you hooked me up to some kind of neurobiological forecasting machine, I promise you I would just pick some totally random thing to think about and your prediction would be wrong. Determinism can't possibly be right."

To respond to that, notice first that the determinist position advanced here does not suggest that you would feel any sense of coercion with respect to waffle-eating. As above, the decision is based on a weighing of much of your experience, and the phenomenology of that frontal / parietal cortical calculation could feel like absolutely free deliberation.

Secondly, there is a complaint here that the determinism of intention implies predictability. This is not true, but the reasons are subtle and important enough to warrant careful explication. Determinism does *not* imply predictability, for one, possibly two reasons: (1) nonlinear dynamics and (2) quantum indeterminacy. We'll explore each of these reasons in depths in the next two sections.

Nonlinear dynamics

Everyone has seen a pendulum in one form or another: a child on a swing, a grandfather clock, or keys on a chain. The motion of a pendulum is very simple: back and forth, back and forth. Of course, regular pendula slow down over time, as friction causes the energy to dissipate, but you can always give a little push and keep it going. Pendula are so straightforward – and periodic phenomena so common and important in nature – that they are usually the first thing you learn in applied mathematics. It's a simple matter to quantify all the important natural laws at work in a pendulum, write down some equations, solve them, and predict quite accurately how the pendulum will behave over time. This is a precise mathematical scheme for what is observationally clear: the pendulum goes tick, tock.

Imagine you have such a pendulum (you can actually do this experiment at home) composed of a rigid rod, like a piece of hanger wire with a weight at the end. You've set up some kind of pivot at the top, so it can swing freely. Assuming everything is stable, and you give this a push, it will display a nice, periodic motion. Now suppose you take a *second* pendulum, with basically the same design, and attach its top to the bottom of the first one: one pendulum hanging from another. The attachment must be such that if you hold the upper pendulum still, the second one is free to swing back and forth easily. This combined system is called, oddly enough, a double pendulum.

You might think that this physical system would be only slightly harder, maybe twice as hard, to understand and predict as the single, solo pendulum. Not so. In fact, it can't be predicted with any power at all. You can recruit the smartest person in the world, equip them with the most powerful computer, and they will not be able to predict the behavior of the double pendulum for even ten seconds. This is surprising, perhaps even a little discouraging.

While there are entire books written about this, we can understand quite simply why this might happen. The basic problem is that the high predictability of the single pendulum depends on a fixed support for the top. If the top isn't fixed but can shake based on the movement of the weight, this causes a feedback force that alters the movement of the weight. That new weight movement then causes the shaking of the top to become irregular, and the pristine mathematics that allowed us to solve the original problem no longer apply. Instead, we can only use what's known as "numerical solutions," in which not much can be done with paper and pencil, but yields instead only to brute force calculations on a computer. While this is fine for many purposes, such as landing on Mars, it is sensitive to measurement error. If we're wrong about the size of the wobble of the top, we'll be wrong about what the pendulum is going to do. Only slightly wrong at first, but becoming worse over time as the error snowballs.

In the double pendulum setup, the bottom of one pendulum is the base of the second, so there is certainly some wobbling going on. The motion of the top pendulum affects the movement of the bottom pendulum. What's worse, since they're attached, is that the movement of the bottom turns right around and affects the top. But the motion of the top is what affects the bottom. This is feedback. And when we combine feedback with realistic (*i.e.* nonlinear) physical laws, we walk right smack into the middle of nonlinear dynamics, where almost every system shows (a) complicated temporal behavior, and (b) a strong sensitivity to measurement error. If we're just slightly off on the measurement, our predictions – which were tenuous in the first place because of the complexity of the system – will become even worse.

One of the best illustrations of sensitive dependence on initial conditions is *the butterfly effect*. A meteorologist uses things like air temperature, pressure, humidity, and velocity to predict the weather. This predictive ability depends on the basic laws of fluid mechanics and thermodynamics, telling us how air moves, what happens when it mixes with other air, how heat is exchanged, and so on. As a thought experiment, imagine you have a meteorological simulator that allows you to take the current readings from air sensors all over the world. You plug those numbers into your software, and when you press go, all the laws of physics that are built in to the simulation start forecasting, minute by minute, how the little pieces of air will change. The process continues as far into the future as you care to predict.

So. You run the simulation and come up with a forecast for, say, the next three days. It predicts sunshine in your city.

Now, go back to the same initial conditions that generated this forecast, and *artificially* change the numbers for some single sensor, somewhere on earth. If the real data showed a small volume of air in Buenos Aires with a temperature of 30° C, change that number to 31° C. Or, if you want to be true to the name, make a small velocity change that would be the equivalent of that little pocket of air experiencing the flap of a butterfly's wings. Then re-run the simulation.

What you may very well see – in fact, exactly this was done by Lorenz in 1961, a meteorologist himself and the man who coined the term – is that by the end of the three-day forecasts, the predictions of the two simulations are wildly discrepant. The one tiny little flap that affected only one tiny little volume of air has ultimately changed the predicted weather for your city, presumably far from Buenos Aires.

How did this happen? It happened in the same way as the double pendulum. The nearly insignificant change for the volume of air you altered has in turn had an effect on the air surrounding it. That air, in turn, affected the air surrounding *it*. On and on the causal chain spread, until eventually, every piece of air in the atmosphere had been affected in some way by the single flap of a butterfly's wings in Buenos Aires. Over the course of an hour or a day, there would probably not much difference between the two simulations, so the causal cascade may not affect the short-term forecast. But if we give the system enough time, a measurement error in the amount of one butterfly-flap will cause incorrect predictions for the future velocity and temperature of air all over the world. Lorenz's finding was of seminal importance in the modern interest in nonlinear dynamics.

All this difficulty arises, moreover, not just for huge computational systems like atmospheric dynamics, but also in the comparatively minimalist case of a couple pieces of hanger wire and weights. Because the atoms comprising the wires are always dancing around a little, it's not really possible to know *exactly* the position in which the double pendulum starts. If you want to predict what it will do, you need to know the position, but you can't measure that *exactly*, to infinite precision; you'd be off by at least the proverbial butterfly-flap. And just as before, that initial error cascades, becoming more and more amplified by the feedback. Eventually – rather quickly, in fact, within seconds – your prediction will be dramatically wrong.

Now, what would be your guess about which object is more complicated: (a) two weights and a couple of pieces of hanger wire, or (b) a neuron? While a single neuron is not as large as the atmosphere, it may very well be more complex due (than the *atmosphere*, mind you, not the double pendulum) to its inhomogeneity and active self-regulation. When coupled to the 100 billion other neurons in your head in the form of circuits that are also self-regulating at a higher level of organization, the system becomes very intricate. Even if you could predict exactly how *one* neuron was going to behave (which you cannot), you would quickly lose that ability when you connected that neuron to another neuron. And another, and another, and another. You should be able to see why, even if neurons were completely deterministic themselves, an entire nervous system could be unpredictable. This is the moral of the story. Neurons, like billiard balls and pendula, are deterministic, *i.e.* completely causal, but unpredictable. The chances that someone in a white lab coat is going to connect you to a machine and tell you what you're going to have for breakfast next Tuesday are nil.

Quantum indeterminacy

The second dissociation between determinism and predictability is quantum indeterminacy. This is not a physics text, so again the technical details will be omitted, but the basic principle should suffice.

Some characteristics of the physical world depend on scale. Surface tension, for example, is a feature of liquids that doesn't seem to have a major effect on our daily life. We can see the effect of cohesion between water molecules in certain circumstances, like the slight curve of water as it climbs up the inside edge of a glass, or the spherical beads of fog that condense on a waxed car. So we can see the effects, but for the scale of a human being, the associated forces are negligibly weak. We can push our finger right through the surface of water in a glass, or push a droplet off the hood with a puff of air. This is because we are big and strong relative to the intermolecular forces of water.

If we were ant-sized, however, things would be different. Go out into a park or a garden sometime and watch how insects interact with water. When you are roughly the same size as a raindrop, the same forces that are meaningless to us become major players in the physical landscape. Water striders live their lives skimming around comfortably on the surface of a pond, their entire weight supported by surface tension. Ants carry droplets of water around like beach balls, no canteen needed.

This is just one example to demonstrate that mere physical size affects the relative importance of different forces. Those features of the physical world most important to ants may be trivially weak for us. For things much smaller than ants, in turn, the world behaves in even stranger and more exotic ways.

Out of the 1920s and 1930s came one of the most profound revolutions in the history of science: quantum mechanics. Driven by scientists such as Dirac and Heisenberg, physical theories of subatomic particles and the forces between them led not only to the most powerful predictive theory ever devised, but also to a philosophical view of nature and causality that challenges the frameworks of philosophers even today – to say nothing of the commonsensical understanding of the world held by nonscientists.

One of the basic ideas destabilized by the quantum revolution was the idea of *position*. This is such a straightforward aspect of our world that no one ever gives it much thought. The world is spatial, and objects have positions. A baseball, at any given moment, has a particular location. My car has a location. Any physical thing has a location, and although I can specify that position in different ways – relative to a wall, or a person, or some other fixed object – there is seemingly no doubt that every object at a particular point in time has a unique position in space.

Before considering how the notion of position was altered by quantum mechanics, let's first think briefly about how we *measure* the position of baseballs and cars. Well, obviously, we look at them. We look at them with our eyes. And why does that work? Because there's light in the world, and that light bounces off the objects in question, and enters our eyes. Fine. But what if it was dark? Assuming we couldn't *create* light to see with, we could always use some other detection mechanism, like sonar. In circumstances where we can't see very far, like underwater, sonar is a fine way of determining position. In that case, we bounce *sound*, not light, off an object. By measuring the time it takes for the sound to travel between us and the object, and then bounce back again, we can determine how far away it is. The similarity between using light or sound to measure position is that we must interact with the object in some way. Either photons or air (or water) molecules hit the object, bounce off, and it is the reflected particles or waves that tell us where the object is. Quantum mechanics drew special attention to the fact that to measure location, we must physically interact with the object being measured.

Position, of course, is not the only spatial descriptor of an object. *Motion* is also important. An object might not be stuck in one position, but rather, be moving. Of course, we can measure motion by simply watching an object or bouncing sonar off of it repeatedly. Different techniques might have different advantages, but irrespective of the technique, we assume that there is an actual location and an actual velocity, and we can come close to perfection in measuring those values depending on the sensitivity of our methods.

In the realm of quantum mechanics, the objects being studied are very small and have very low mass when compared to the objects we interact with everyday. An electron and a baseball, for example, have about the same mass ratio as a mosquito and the planet Earth. (Actually, the difference is 100 times *more* pronounced in the electron/baseball comparison). As a result of the small masses, forces negligible for objects the size of people become quite significant at the quantum scale. The flight path of a mosquito can clearly be influenced by forces subtler than those affecting the trajectory of the entire planet.

Suppose you have captured an electron in a box and you want to know precisely where it is. What do you do? You look at it. Never mind for the moment that you can't see it with your naked eye...let's say you have a special microscope. The microscope emits photons, some of them hit the electron, and when they come back to the microscope, the flash of light lets you know where the electron is. As straightforward as that seems, there's a problem. Light, because it is a form of energy, has *momentum*. When the photons in the light hit the electron, it's like a BB hitting a mosquito. By the deflection of the BB, you can get some idea of the position of the mosquito, but at the same time, you alter the *velocity* of the mosquito. If you knew the velocity prior to the position measurement, you don't know it any longer.

It turns out that there is no way around this problem. There is no way to measure the position of an electron, or any other object, for that matter, without affecting its velocity. Conversely (though for slightly different reasons) there is no way to measure velocity without affecting position. Measurement of the two features is conjoined into a zero sum game. The better one measurement gets, the worse the other gets. For large objects like baseballs, the momentum of the light that we use is very small compared to the momentum of the object, so the velocity perturbation is negligible. But for small objects like electrons, it's *not* negligible. It can be significant enough to seriously impair our ability to understand the state of the world.

Much has been written about the philosophical consequences of this situation. While there are still many open questions, one point on which there is consensus is that the proper mathematics to use when discussing quantum mechanical systems is the mathematics of probability. That is to say, if we wish to predict the results of an experiment, the best – presently the *only* – thing that we can do is to provide a probability distribution for all the possible results. We can't predict specifically what will happen, *e.g.* what precise velocity we will measure, but we can give a probability distribution showing the likelihood of all possible velocities.

Probability, of course, is not certainty. If the quantities we wish to predict are associated with particles in your brain, this means that the future state of your brain cannot be known with certainty. Practically speaking, the prospect of using quantum mechanics to predict the future state of an object as large as even a single neuron is preposterous. Calculations of that magnitude may well be permanently inaccessible, but even if not, they will still be probabilistic rather than point-value predictions. Some of those seeking a special place for human consciousness have seized upon this fact and consoled themselves that they are not, in the end, billiard balls. They are wrong to do so, however, because the physics which make a neuron predictable are equally applicable to *every* macroscopic physical system. There's nothing special about brains that makes them more unpredictable, in a quantum sense, than cheeseburgers or billiard balls.

Exactly where this indeterminacy comes from is, again, a subject of much interest. One view, arising from the historical evidence given above, is that it is deeply connected to the idea of observation and measurement. At the end of the day, observation implies interaction, and interaction is never completely passive. Even if there is no consciousness (whatever that is) in the role of observer, any other particle in the universe that "wants" to interact with a given particle is in effectively the same position: if two particles collide, the outcome of the collision depends on each particle "finding out" what the position and velocity of the other is. This being the case, *i.e.* given that it is impossible even in principle to determine position and velocity simultaneously and with infinite precision, can we even say that particles *have* unique positions and velocities? *I.e.* is there is an inherent fuzziness about nature? Or would we be justified in believing that nature is inherently crisp, albeit in a way that has no measurable or practical consequences? These are deep questions that should not be contemplated without a much better grounding in the mathematics.

Although some writers (*e.g.* Penrose & Hameroff) have probed these depths in search of some sort of acausal *deus ex machina* where free will could reside, no such effort has been received by scientific consensus as even a remotely likely possibility. Penrose's position in particular has been convincingly rebutted.[24] Moreover, the "nature is fuzzy" possibility, also known as the Copenhagen interpretation, is not favored by philosophers of physics, who tend to place more confidence in the "nature is crisp" possibility – and in this latter case, there is no room for acausal agency. Acausality is precisely what would be required for free will as I have defined it. If nature is completely causal, free will as it is generally understood is impossible (and in advocating this view, I identify myself as an *incompatibilist*). If nature is not causal, then the indeterminacy lies deep in the fabric of elementary particles, and it becomes very hard to see how human brains, or brains of any animal, are any different with respect to their particulate composition than other massive objects, like eggplants. Why the quantum mechanics of brain tissue, but not eggplant tissue, would give rise to free will would have to be addressed by its defenders. Penrose & Hameroff argue that this involves the microtubules inside individual neurons, but again, this has been convincingly refuted and is not regarded by even a small minority of neuroscientists as a likely basis for any macroscopic aspect of psychology.

We have now reached the end of the discussion on the difference between determinism and predictability. The take-home message is that a system can be completely *causal* but still *unpredictable* (à la nonlinear dynamics) and that nature itself *may* not even be causal. The brain – although not just the brain...the entire universe in fact – is such a system. Even if free will does not exist (my argument is that it does not), even if your mind is just a mechanistic slave to the underlying physics (and it seems to be just that), this does *not* mean that anyone will ever be able to predict what you are going to have for breakfast next Tuesday. It seems more likely than not that they will never have that ability.

The real source of behavior

Nonlinear dynamics and quantum indeterminacy are both appeasements for the complainant at the beginning of the chapter, who was upset that someone might be able to wire him up to a machine and predict his actions. As we've seen, that fear is unfounded. The complainant also argued, without much support, that it just didn't feel like their will was pushed around by physics. We can concede that it doesn't *feel* like it is, but how much weight does this feeling deserve in a scientific investigation? Does it *feel* like $F = G\frac{m_1 m_2}{r^2}$? Does it *feel* like $H\Psi(\mathbf{r},t) = i\hbar \frac{\partial}{\partial t}\Psi(\mathbf{r},t)$? There is no good reason why intuition about minds and brains should be taken any more seriously than intuition about other natural processes. Complex and counterintuitive laws regulate even such apparently simple physical systems as three-body gravitation and lone electrons; why should a vast collection of self-regulating neurons be any different? All right – but if your mind is a passive consequence of physics, what's the argument? If our thoughts and actions don't come from us, from our will, where *do* they come from, and why don't we act like billiard balls?

The basic instructions for how to build a brain are encoded in your DNA (the "nature" in the nature / nurture debate). The instructions required to build different types of neurons, grow processes and synapses connecting them to their functional partners, and construct various transmitters and receptors – these instructions are all contained in the genome. Beyond that, however, the microcircuitry of the brain, *i.e.* the specific cell-by-cell, dendrite-by-dendrite connectivity, is extremely dependent on experience ("nurture"). In one sense, this is obvious when you look at an infant. It takes years for an infant to develop coordinated motor skills, perceptual sophistication, cognitive competence, and reliable planning. The circuits supporting these functions are composed of trillions of connections between neurons. While DNA does encode a great deal, it cannot encode anywhere near enough to specify the precise details of how all the connections in your brain should be formed. Moreover, DNA does not "know" what kind of world it is going to encounter, and therefore what sort of microcircuits it should build. Is it going to build a brain for a pianist, a physicist, or a plumber? How could the circuit for binocular depth perception be built when DNA doesn't "know" until growth – which is affected by diet – how far apart the eyes will be?

It is therefore quite essential that the details of brain microcircuitry be fine-tuned by experience. And experience, we must remember, is never the same for two people. Even identical twins who live their whole lives together have sufficient differences in experience to produce two very different neural microcircuitries. One twin happens to be looking up when a bird flies overhead – the other is looking down. One sleeps while the other helps make cookies. Every one of the millions of such events has slightly different effects on each twin. And the process builds upon itself. If you make cookies one day, you'll react a little differently next time they're made, and that will further emphasize the difference between your experience and the twin who slept. Feedback and nonlinear dynamics again.

Environmental factors such as nutrition also fall under the "nurture" heading. If one twin has a peanut butter and jelly sandwich for lunch when the other has grilled cheese, different nutrients make their way into the bloodstream and are delivered to the growing brain. One particular cell gets a protein instead of a sugar, which enables it to grow a tiny dendritic process. That anatomical change alters the environment for neighboring neurons, which now change their behavior, and so on.

All of these developmental and experiential differences mean that two people never have the same brains, even with identical DNA and "identical" lives. Anatomy, however, is not the end of the story when it comes to behavior generation. One must also understand the function that the brain performs and how it does so.

One way to think about the function of the brain is as a big input / output device that continuously runs a three-phase cycle. (1) Process input from the senses, (2) select an appropriate behavior, and (3) generate the behavior and return to Step One. The exact nature of what is computed certainly depends on what comes in, but probably the single largest determinant of individual behavioral uniqueness is Step Two, which is what we ultimately call the decision-making process. All the developmental and experiential history of the individual comes into play here, including the details of all those microcircuits, as the pros and cons of different potential behaviors are weighed. For each action we might do, the brain forecasts the future world states likely to result, grades those futures based on desirability, and executes the plan with the biggest payoff.

Although it is possible for the brain to select behaviors that are in conflict with the best interest of the genes that built it (*e.g.* suicide or voluntary childlessness), the genes have done a remarkable job of producing a device that, in general, calculates payoffs that are indeed Darwinian. Although most of us don't think about everyday events in terms of Darwinian fitness, a little introspection can demonstrate that it is very often the ultimate basis for our decision-making. For example: what's for lunch – cheeseburger or chicken caesar? We might think to ourselves, "Hm…cheeseburger: so tasty, but bad for you and the environment. Chicken caesar: lighter, more nutritious, but then I'll be hungry later and probably just eat chips or cookies, and then be worse off." The ultimate calculations here are about caloric content – a very important feature of food in the archaic environment – and the tradeoff with long-term health risks. Dying is rarely advantageous on

Darwinian calculators, so the actual risk of a single cheeseburger *vs.* a couple trips to the snack machine need to be carefully deliberated.

Neurobiologically, what's happening during this process is that your frontal lobe and emotional centers in the amygdala are running a winner-take-all election between behaviors representing the two food choices. One of those candidates ultimately wins the competition and is automatically sent to the motor cortex to initiate movement. During the time that votes are being tallied, you are unsure about what you're going to do. If someone asks you what you're doing, you'll say, "I'm trying to make up my mind." That's true in a way – your mind is in the process of being "made" – but there's really no need to posit an "I" that runs the show. It would more parsimonious just to think of it in the following way. The current state of the world is perceived and processed by your sensory cortex. The decision-making part of your brain is informed that there are two choices: cheeseburger and salad. Long-term memory is consulted to determine prior experience with these foods. Any nauseating memories? Is the flavor appealing? How have we typically felt an hour after eating? How hungry are we now? What have we eaten recently? Are we short of any specific nutrients? All of this happens in seconds and often *sub tabula,* without significant awareness. Eventually, taking everything into account with whatever level of analysis is appropriate, a selection is made, and off we go. There is no overseeing superego directing the process. There is just the frontal cortex, memory, emotion, and a payoff calculator, all flowing through a series of mechanical steps based on the current state of the world and your prior experiences. The "I" that you feel is nothing less, or more, than all of these systems interacting.

Like every one of the component systems that comprise it, from ribosomes to mitochondria and the Krebs cycle, your brain and the consciousness it produces are machines. The brain is a very fancy, biological, recursive, adaptive, self-aware machine, made out of lipid bilayers and sodium channels, but a machine nonetheless. It is a wonder to behold. To many who understand this, it is in fact far *more* wonderful than the unsupportable positing of an immaterial will, which itself is incomprehensible and explains nothing. This is, in miniature, the difference between the naturalistic and the supernatural worldviews.

*   *   *

If the process of making a decision is as mechanistic as I have portrayed it, with no free will at its center, why doesn't it feel that way? Because the way it feels, frankly, is that the aforementioned calculations are done – at least, in some situations – but it is *me* who is doing them. Not that I *am* them.

Just as a thought experiment (no pun intended), consider this. Can you imagine what it would be like to be something else? If you say that it doesn't feel like you are a bunch of calculations, this implies that you know what it *would* feel like to be a bunch of calculations, and the way you feel isn't that. But can you really imagine such a thing? To take the original example from Nagel, what is it like to be a bat? One of the difficulties in answering a question like this is that it will always be *you* doing the imagining, and therefore, how can you extract – or, really, subtract – yourself from the situation? If the one universal in imagining being something is you, and you come complete with memories, emotions, preferences, abilities, idiosyncrasies – including those associated with how you go about imagining something – then how could one compensate for all those things? And if they are all based on an even deeper substrate of biological computation, it seems even less likely that you would be able to subtract *that* out of the equation. Anyway, if all these things were removed from the imaginary scenario of being something else, *you* would, by definition, be gone. There would be no one left to finish the job! It may just be logically impossible to imagine not being you.

Let's take a simpler example. Try to imagine being you, but in an alternative universe where, instead of action potentials being generated by sodium ions, they are generated by lithium ions instead. Everything else the same. Got it? And? Does it still feel like you?

We really are not good at imagining what it would be like to be something else. We can imagine being mostly like ourselves, but *slightly* different – perhaps if we were just like this except that we liked Brussels sprouts. Or just like this, but an airline pilot. But such things are really changes of circumstance, not of underlying identity. When that identity descends into the foundations of the hardware that supports our consciousness, it gets downright impossible to imagine. Mustn't we admit that, at the end of the day, we really don't know what the consequences would be if our decisions really *were* just a bunch of frontal cortex / amygdaloid calculations? What would it feel like if they were *temporal* cortex calculations? You don't know, and neither do I, and that means we really should recuse ourselves from the case. We just are not good enough at this kind of work to have a basis – a purely introspective basis at that – for determining what the physical substrate may be of our decision making process. None of this argues persuasively for the frontal cortex process outlined above to be *the* process – that evidence originates from cognitive neuroscience – but it does defuse the objection that the frontal cortex can't possibly be the source of our decision making because it doesn't feel that way.

There is another way to think about the illusion of free will, and that is the evolutionary approach. Many of the features (but not all!) of an organism have adaptive value. Clearly, what I am calling the illusion of free will is one of our features. Even I, who don't believe in free will, *feel like* I have it. Is there some evolutionary benefit to that feeling? Why did it evolve?

Perhaps the best answer to that is to consider the alternative. Suppose that somewhere in the course of evolution an animal arose that felt like it had no control over its own actions. It really felt, deep down, that the universe was deterministic, and everything it was going to do had already been written in the stars, so to speak. Would this animal be better off or worse off than other animals, who believed that they were masters of their own destiny?

Before answering, let's examine the question a little more carefully. What does it mean for an animal to feel like it has no control over its own actions? This means that its emotional systems (this is the "feeling" part) are disengaged from motor program selection (this is the "action" part). To actually feel like you can just sit back and watch the show of your own behavior is to remain emotionally neutral over the prospect of any action or inaction, however imminent, however central to survival. But we have already discussed, and it's intuitive anyway, that emotional valence is a critical part of motor program selection. Emotion is, in effect, the voice of Darwinian advantage. If something feels good, that's evolution telling you that as far as it knows, you should keep doing it, that it's good for survival and reproduction. So if your emotions are disconnected from motor program selection, you are in very bad shape. Your frontal cortex is essentially free wheeling, selecting behaviors more or less at random, with no computation to determine which is likely to yield the greatest reward. An animal with this circuit is probably not going to fare very well, and the gene that produced this critter will slide into oblivion.

Note, however, that this does not mean that the feeling of free will is necessary for survival. Bacteria are very, very successful, but we have no idea whether they feel like they are free. Most betting types would give good odds to the claim that they don't. But they still need some calculator – emotion – that tells them which things feel good and which feel bad. At the core of any behavior-selecting organism, then, must be some analog of pain and pleasure. Whether higher layers of reflection exist in addition to that basic level is another question, and the reflective feeling of being free to direct the course of one's own thoughts and actions is just such a higher layer. Although most of us would reasonably doubt that a staphylococcus is reflective in that way, such skepticism becomes harder to sustain with respect to megafauna like mammals and birds.

For the last time: on a personal note, I really, *really* do not think that I have free will. But notice, I said that's what I *think*. When it comes to feelings, I very much *feel* like I am free. But this is not saying much – feelings are not often based on logic or evidence, and per the above argument, my feeling of freedom simply means that my emotional centers are connected to my frontal cortex for the purposes of motor program selection. The fact that I am alive today should indicate that that was probably true anyway.

Death and regeneration

In non-naturalist worldviews, particularly theistic worldviews, free will is usually tied in one way or another to the soul. The soul, of course, is the immaterial (or material but somehow magical) thing that contains our personality. This is an important construct for anyone who wishes to find something divine about humanity, because it's quite clear that in organic terms, we are rather ordinary primates. Even the brains that give rise to our extraordinary intellects are, in biological terms, merely quantitatively different from those of, say, chimpanzees. And their brains are, in turn, only quantitatively different from macaques, and so on. Yet the religious presupposition is that there is something special about the human personality, allowing it to transcend biological mortality via a continued existence in an ethereal plane like heaven or via reincarnation and a return to earth. Because such beliefs are so emotionally important to believers, the life after death question really must be addressed by any ambitious naturalist.

First, the bad news. Naturalism says that with respect to life after death, there isn't one – at least, not as that idea is conventionally understood. There is no everlasting utopia for your soul, not least because there is no such thing as an immaterial soul. The soul, as we generally use the term, is your personality, your values, your memories, your will, your self. And yet, all of these things are demonstrably products of the brain: changes to the brain, for better or for worse, can affect every one of them. The brain is a completely embodied thing, and it is a long-known difficulty with dualism that interactions of an immaterial soul with a material brain are irreconcilably non-naturalistic.

This conclusion of naturalism, that there is a very low likelihood of life after death, is unfortunate. Not even scientists, in their perverse and godless ways, want to believe that an errant bus could end their lives senselessly and prematurely. Luckily, there is *some* consolation about death in the naturalist worldview, but before getting to that, a few words about death *per se* may be helpful.

Strictly speaking, the finality of death doesn't require a naturalist defense because, as with all things, the burden of proof rests with the person making the claim. Naturalism merely recapitulates what we can all directly observe, and therefore doesn't really make any claims at all. When an organism dies, it stops behaving as it did, ceases consumption and homeostasis, and its body disintegrates. If we wait long enough, we can see for ourselves that death eliminates all the essential personal characteristics of the former life. Its physical presence and its dynamic properties simply disappear. Naturalism just repeats what we see; it points a finger at nature and says, "Look."

This is only half true, however, and this is where the good news comes in. Don't get your hopes up. This is not the pie-in-the-sky scenario of religion's life after death claim, but it is profound and glorious in its own way.

Science recognizes an important distinction between (a) the material from which a thing is made, and (b) the way that material is organized. You can take a fixed amount of a single material, organize it in two different ways, and get two completely different results. Not subtly different, but *deeply* different. For example, graphite (pencil lead) and diamond are both made completely out of carbon. There are no elements other than carbon present in either material. If you had a pile of carbon atoms and sufficient technical skill, you could divide the pile in half and build a pencil lead from one side and an engagement ring solitaire from the other. These two materials look different and act different, but their differences lie purely in how their atoms are arranged.

Judging from its manifest of component materials, the phenomenon that we call "life" is nothing special. Carbon, hydrogen, oxygen, nitrogen, …these are all common elements, and one can assemble an enormous variety of boring, inert chemicals from them. What makes life *alive* is the particular way the atoms are arranged into molecules, and the relationship that the molecules have with each other. If you remember anything about cellular biology and go out into your garden or the woods, you can achieve moments of great astonishment solely by contemplating the multitude of intricate relationships that exist between the molecules of living things. Inside even the tiniest single-celled organism is a vast economy of exchanges, electron transfers, phosphorylations, oxidations, and catalyzed reactions. And it all happens at lightning speed with almost no mistakes and no intelligent supervision. Again, it's a wonder to behold.

There are many ways to die, but in the end, pretty much every one of them has the same result. Death hampers the working relationships between molecules, which leads to a cascade of structural change – change occurring at the level of individual molecules. If sufficiently many of your cells are deprived of oxygen or nutrients or poisoned by their own waste products, the huge interdependent factory that is your body starts to fall apart, and eventually stops. Given the number of things that *could* go wrong, life is impressively robust, but sadly, some perturbations are just outside its operational limits, and it breaks down.

After death, where do those intercellular and intermolecular relationships go? Nowhere. They simply disappear, like your lap disappears when you stand up. Unfortunately for you, it was in those molecules and relationships that your personality, your feelings, ideas, and memories – what some would call your soul – lived.

Functionally, you're dead. Materially speaking, however, judging just by the list of elements and their relative amounts, not much has changed. Your body still has the same carbon, hydrogen, oxygen, and so on that it had in the moments before everything shut down. Atoms are very sturdy objects (molecules and cells are much more delicate). Every atom in your body was formed in the thermonuclear furnace of a star somewhere, probably several billion years ago. For all the countless eons before you were conceived, every one of your atoms was carousing around, perhaps being blasted out of a volcano on the Pacific Rim, or drifting around in the high atmosphere. Some of them got the smackdown of a lifetime when the asteroid that killed the dinosaurs landed in the Yucatán. Still others spent some time in the nervous system of a prehistoric fish. In fact, there is a wonderful calculation estimating that each person living today has inherited 200 billion atoms that once made William Shakespeare![25]

The point is that atoms are forever. Ok...perhaps not forever, but a long time. What happens to them? After you're done borrowing them for your teensy little lifespan, most of them will gallivant around the biosphere for another couple hundred million years. If you're lucky, some of your carbon will be used in the pencil of a school-bound sentient cuttlefish, or your hydrogen will end up in a mid-day cheese plate for an evolved meerkat. Some of your constituent elements may be quiet, stay-at-home types, lingering in a block of coal for 50 million years, while others flit from fern to fungus to firefly, rarely leaving the biosphere. Eventually, the meerkats and fireflies too will die, and the atoms will move yet again, and again, into the far distant future, several billion years from now. It is somewhere in that timeframe that our sun will run out of fuel, engulf the earth and other planets, and the whole collection will turn into some kind of collapsed, dead dwarf star trapped on a slow gravitational waltz through the cosmos. Several billion years after *that*, the atoms you see this very moment in the skin of your knuckles may be lost in the singularity of a black hole at the center of the Milky Way galaxy. And for what happens after that, you'll have to ask Stephen Hawking.

This is the kind of immortality that naturalism has to offer. It is different from the theistic brand, because (a) your consciousness doesn't continue, and (b) we do not know the fate of the universe, so the true end, if there is one, of your material body (or, rather, of the material *making up* your body) is unknown. There's no way to know precisely what your atoms did before you got them, or what they'll do later, but one thing is certain. The thread that weaves all those histories together – and there's one thread per atom, or about $7 \cdot 10^{27}$ of them – makes the story of your atomic origins and legacy totally unique. It would also make a *great* – although rather long – movie. Trumping all of that is the chief virtue of this naturalistic account: namely, whatever consolation you glean from it is empirically defensible. What more could you reasonably ask of a philosophy of life (and death) than that it be demonstrably correct?

Our relationship with the natural world

One way of thinking about humanity's place in the world is as one of privileged superiority. This is the Genesis position. God created us and set us apart, instructing us to use nature for our own private comforts. Even a fairly relaxed interpretation of the Old Testament – to choose just one representative text – spells out this position clearly. Biblical interpretation is notoriously subjective, but even moving into the more widely embraced New Testament, it is probably fair to say that most

Christians, even if they believe in evolution, believe that God played and possibly still plays some role in the creation and betterment of our particular species. Jesus Christ did not die for the sins of hedgehogs, after all. You almost *have* to believe that God plays an ongoing role if humanity is to have any sort of privileged position, whether it be creation in the likeness of God, possession of a soul, admission to heaven, or something similar. Many, if not most, religions say that people are special in a way that no other animal is.

The naturalist position, in contrast, says that although our species may be privileged by virtue of certain intellectual abilities, this is not by entitlement or by destiny. Rather, like all other species, *Homo sapiens* is the product of undirected natural selection. We *happen* to have evolved a set of abilities that puts us in a unique position of control over many aspects of our environment, but this is not because some higher being willed it to be so. Our special strengths are products of adaptation over the millennia, just like a giraffe's neck or a cheetah's speed.

Some of the theist / deist persuasion find the naturalist claim insulting. This is not surprising. If you have grown up believing that the universe has a special interest in you personally, that some omnipotent all-seeing power is watching you and trying to clear the way for you, making sure everything turns out ok, it must be depressing to contemplate a cold, empty, nihilistic alternative that says, "You are not special. No one is watching. Everything might not be ok. The universe is big and unforgiving and you are just a peculiar arrangement of protoplasm." For some, this feels like going from the head of the class to the bottom of the dumpster.

What I want to do here is to speak to those who feel this way and see if I can replace the sense of loss with something of equal or even greater value.

First, consider that naturalism does not claim that humanity is at the bottom of the heap. By noting the biological meaninglessness of "entitlement" and "privilege," naturalism merely puts us shoulder to shoulder with our fellow creatures. Not ahead, not behind, but right alongside. So this idea of going to the bottom of the dumpster is really unjustified. Don't feel like you're the unwanted child...you're just not the heaven-sent incarnation of the divine that you thought you were.

Moreover, this is just in a biological sense. You are still *very* justified in a sense of pride about humanity's abilities and accomplishments. Our species' natural history is an extraordinary tale of cognitive skill, social cooperation, and above all, symbolic and linguistic superiority. All of these talents have given rise to an unprecedented strength in manipulating our environment, our meteoric rise to power and comfort, and our astonishing intellectual and artistic achievement. Chimps and dolphins lack both antibiotics and Public Television. Whether all our skills will prove adaptive in the long term is another question.

On a more grandiose and eerily circular note, think about this. Naturalism says yes, our bodies are the products of undirected evolution. And yes, so are our brains, and hence, our minds. But since it is precisely our minds that have learned so much about the structure and function of the world, there is a certain closure in which the universe, through us, is coming to understand itself.[26] Imagine the billion-year process beginning with a swirling plasma cloud, which condensed into a proto-galaxy, which gave way to stars, to our solar system, to our planet, to life, and to our sentience. Although that arc is only one of many on the evolutionary tree, we are justified in feeling special about our position as explorer and discoverer. We act, if you will, on behalf of the universe itself. And can you imagine how exciting it would be to discover how alien civilizations have, in their own way, participated in the same cosmic scheme? How have they arranged the natural patterns they observed? Is our theoretic scheme of modern science the only one possible? What does alien *philosophy* look like?

It may be impossible to convey in words the vertigo, the sublime braid of emotion and idea, that this view entails. For a naturalist, a walk in the woods is a confrontation with the divine, small d. To contemplate the abundance of life, the vast harmonies between interacting systems, the total rootedness of our own history in the context of the entire scheme is to make a personal connection with something that is at once ephemeral and eternal. Some religions are emotionally appealing because in their best moments, they exude a feeling of total acceptance, as if God knows your soul and you are loved. Nature, for a naturalist, can be the same way. All around us are brothers and sisters, close cousins, distant ancestors, in incredible diversity and buzzing with life, and we are welcome among them.

Unlike church, however, there is no rank in the forest. There is no God before whom you must prostrate yourself. There is no jealousy, no exclusion, and no judgment. There are no intermediaries, and you don't have to wait until you die to be connected to eternity. It sounds a little like pantheism, but it's simply molecular biology. Written into the DNA and the fossil record of the biosphere, the entire story of your family tree is written, complete with births, deaths, marriages, and children. *That* is the naturalist's holy book, written in A, T, G, and C.

Those who accept on faith the religious claim that humanity is *not* inextricably bound to the natural world are barred from participating in this grand and beautiful cycle. To cling to entitlement and privilege is to demand separation of our species from the world, to disown our biological heritage. You can be religious, but you can't be religious and fully participate in life at the same time: the price of admission to heaven is loss of membership in the society of earth, and kinship with the creatures who inhabit it.

Naturalists are not trying to abolish the emotional comforts of religion. We just want to replace the feelings of belonging and grandeur with something justifiable, something that doesn't demand that we take leave of our senses and rationality.

Evolution

If there is a single idea that defines our relationship with the rest of the biosphere, it is evolution. Because of its monumental importance and staggering simplicity, it is dismaying in the extreme that *anyone* in the developed world doesn't understand it. And yet, many, if not most, don't. Part of the reason for that is bad education, but part is religion. Since creationism's PR re-labeling as Intelligent Design, many of the old religious misunderstandings of evolution have resurfaced. Biology in general demonstrates such strong ties between humans and other living things that it is a natural target for an ideology that seeks to widen the distance. While this book is not the place for a full exposition of molecular or evolutionary biology, and others have done it better, it is such an essential part of the naturalist's worldview that I do feel obliged to make a point on one key issue.

One of the most prominent advocates of Intelligent Design, Michael Behe (1998) introduces the concept of "irreducible complexity" as a central argument against evolution by natural selection. In particular, he argues that the gradual accumulation of small changes, each conferring a small selective advantage, cannot account for the presence of natural mechanisms in which rich interdependence precludes gradual and linear evolution. In other words, Behe claims that some natural systems are so complex, and their parts so interdependent, that they must have appeared all at once – and such sudden appearances are more easily understood via the actions of a goal-driven intelligent designer than by gradual Darwinian mechanisms. To introduce his argument, let's first consider his example of a human-engineered mousetrap and its status as an example of an irreducibly complex system. Behe writes:

> Consider the humble mousetrap [(Figure 3-1)]. The mousetraps that my family uses in our home to deal with unwelcome rodents consist of a number of parts. There are: 1) a flat wooden platform to act as a base; 2) a metal hammer, which does the actual job of crushing the little mouse; 3) a wire spring with extended ends to press against the platform and the hammer when the

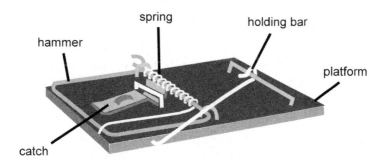

Figure 3-1. Mousetrap.

> trap is charged; 4) a sensitive catch which releases when slight pressure is applied, and 5) a metal bar which holds the hammer back when the trap is charged and connects to the catch. There are also assorted staples and screws to hold the system together.
>
> If any one of the components of the mousetrap (the base, hammer, spring, catch, or holding bar) is removed, then the trap does not function. In other words, the simple little mousetrap has no ability to trap a mouse until several separate parts are all assembled.

Because the mousetrap is necessarily composed of several parts, it is irreducibly complex. Thus, irreducibly complex systems exist.

Obviously, mousetraps are not naturally occurring devices, but Behe goes on to describe in some detail the molecular machinery of cilia, one of the microscopic propulsive organelles used by cells. He makes the case that cilia are also irreducibly complex, in the sense that the removal of any component abolishes the propulsive function. Therefore, he argues, cilia could not arise from anything other than a designer:

> But since the complexity of the cilium is irreducible, then it can not have functional precursors. Since the irreducibly complex cilium can not have functional precursors it can not be produced by natural selection, which requires a continuum of function to work. Natural selection is powerless when there is no function to select. We can go further and say that, if the cilium can not be produced by natural selection, then the cilium was designed.

Thus, according to Behe, irreducibly complex mechanisms cannot have functional precursors. This is an important but incorrect claim, and one which, when understood fully, illustrates a subtle point about evolution and fitness landscapes. Indeed, it addresses his central complaint and shows how natural selection *can* produce irreducibly complex mechanisms. To see how, let's return to the mousetrap. First, I will assume that Behe would accept the following definition:

> Irreducibly complex (def.): A multi-part mechanism is irreducibly complex when the removal of any single part destroys the function of the whole mechanism.

Let us grant that the mousetrap of Fig. 3-1 is indeed irreducibly complex according to the above definition. Barring some hairsplitting subtleties, it is fairly obvious that the removal of any one part of the mousetrap will destroy its function as a killer of mice. Indeed, even a change in the functional relationships between the parts will destroy the function of the mousetrap as a killer of mice. If, for instance, the holding bar were shortened in such a way that its end no longer fit into the tab on top of the catch, then it could not hold back the hammer, and the entire design of the trap would be foiled.

Consider, however, the modified device shown in Fig. 3-2. In this design, exactly such a change has been introduced: The shape of the hammer has been changed so that it no longer extends over the catch. It should be clear that this version of the device would not be likely to work at all well as a mousetrap. When a mouse perturbs the holding bar by moving bait on the catch, this will indeed trigger a spring release of the hammer, but it is too short and will not strike the mouse. So far, I agree with Behe: an alteration to just one part has destroyed the mouse-killing function of the mechanism.

Note, however, that this modified device can be seen as having a different (non-mouse-killing) function. Namely, it is a catapult (mousapult?). The undershooting hammer is aligned with the proximal lever arm of the catch. A projectile placed on the catch will be vaulted some distance when the spring release mechanism is triggered.

While Behe is correct that a change to an irreducibly complex mechanism can destroy the function of that mechanism, the same change can, at the same time, create another *different* function. In this case, the mouse-killing function of the device has been destroyed, but the catapulting function has been created. Or, if one imagines going in the opposite direction, from the device of Fig. 3-2 to that of Fig. 3-

1, an irreducibly complex catapult is destroyed and a fully functional mousetrap is created in its place.

For many purposes, the mousetrap / catapult is an instructive and useful example in our discussion of irreducible complexity, but we must at this

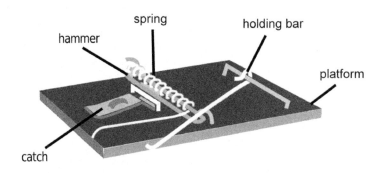

Figure 3-2. Modified mousetrap / catapult. The only change from the device shown in Figure 3-1 is that the hammer of Figure 3-1 has been shortened.

point make note of an important analogical weakness. Seen from the perspective of goal-driven mouse killers, we judge the effectiveness (or "fitness") of the Fig. 3-1 device by its ability to kill mice. From that perspective, the change in the shape of the hammer has completely destroyed the fitness of the device. The catapult does not kill mice.

Most biological entities, however, operate in larger fitness landscapes wherein the loss of one function and the gain of another is not necessarily fatal. To interpret the change in function of the device from a more realistic and in which – who knows? – the ability to catapult small projectiles might very well confer a selective advantage outweighing the loss of mouse-killing ability. Indeed, gene homology studies have demonstrated that at least in some cases, structures originally used for one purpose have, through modification, come to be used for something quite different – this appears to be the case in the evolution of insect wings from the gills of their aquatic ancestors.[27]

The possible gain in fitness resulting from a functional shift is even possible, in principle, for Behe's more serious and literal discussion of cilia. Behe states:

> Cilia are composed of at least a half dozen proteins: alpha-tubulin, beta-tubulin, dynein, nexin, spoke protein, and a central bridge protein. These combine to perform one task, ciliary motion, and all of these proteins must be present for the cilium to function. If the tubulins are absent, then there are no filaments to slide; if the dynein is missing, then the cilium remains rigid and motionless; if nexin or the other connecting proteins are missing, then the axoneme falls apart when the filaments slide.

To propose just one example, an organism possessing all these components except dynein would, according to Behe, produce a non-motile, rigid cilium. While useless as a propulsor, this structure might still increase evolutionary fitness to a single-celled organism, perhaps in the same way that external spines confer a defensive advantage to sea urchins – namely, as a purely mechanical defense against predators, or as an internal organizing structure.[28] Or, rigid cilia could aid in the entrapment and filtration of food particles, which would again be evolutionarily advantageous. In either case, we need not conclude that it is only the fully-formed, irreducibly complex structure that conveys evolutionary advantage. Other "reduced" structures can also convey advantages, albeit in different parts of the fitness landscape.

As a final matter of precision, we should address the possible complaint that the modification introduced by the Fig. 3-2 device does not actually correspond to the working

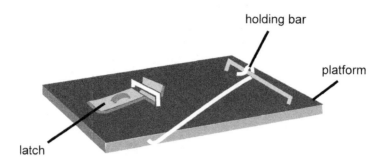

Figure 3-3. Simple lock.

definition of "irreducibly complex," inasmuch as it modifies a part rather than removes it. To that point, we consider the device of Fig. 3-3, which does remove parts, and can also be interpreted as having possibly useful functions, again, in other domains of a more inclusive fitness landscape. In this case, the reduced device functions as a lock.

In effect, Behe's error is that he evaluates natural mechanisms teleologically, as if the function they perform today is the only function of importance to the organism. While Darwinian gradualism does suggest that in order to sustain a presence in the gene pool each mutation must confer an advantage, it does not claim that those advantages all derive from the fitness conferred by a single biological function. Instead, successive mutations may qualitatively change the function performed by the mechanism, conferring advantages from other areas of fitness space. Mousetraps do not have to evolve piece by piece, each generation better than the next. They can emerge, fully-formed, as completely dysfunctional, mutant catapults.

Irreducible complexity may very well be a feature of natural mechanisms, but contrary to Behe's claims, it can emerge from blind (*i.e.* non-goal-driven), gradual, selective processes. It does not imply intelligent design.

CHAPTER 4

The Science of Philosophy

E ven if faith-based religion were diminished as a guiding force, a gaggle of secular philosophers would still do their best to bring structure and direction to human experience. That tradition has existed both within and without religion since at least the Pre-Socratics. But "If not God, then Philosophy" is not good enough, since it merely replaces one metaphysics with another. For a naturalist, metaphysical precepts are just as inscrutable as those based on faith. Where do metaphysical precepts come from? They come from the hands and mouths of metaphysicians. And where did the motions of those hands and mouths come from? From the brain, and you know the rest.

It is an ambitious undertaking to explain in naturalistic terms even a tiny minority of the most important philosophical debates, but that is exactly what we are going to attempt. It's such a tiny piece, however, that all we can reasonably hope to accomplish is something in the spirit of what mathematicians call an "existence proof," *i.e.* a bite-sized demonstration that the project, if undertaken more thoroughly, is plausible. Selecting just a few philosophical topics, I will try to show how each might be explained as originating not from metaphysical truths that lie beyond space and time, but from idiosyncrasies in the evolution-driven functions of the human brain. For this exercise, it is not particularly important whether the philosophies selected are well-regarded by contemporary scholars. The mere existence of an idea is the datum to be explained.

Ultimately, biology causes philosophy, whether it be religious or secular. If the most thorough study indicates that physics or mathematics is at the bottom of it all, then let's go to the bottom and build our foundations there. And even if metaphysical investigation is what led us to believe that physics or mathematics is at the bottom, then we need to use that conclusion to slowly deconstruct the philosophical scaffolding and stand, for the first time since the whole thing began, on the earth.

Sartre and Hegel

In a classic example from *Being and Nothingness*, the twentieth century existentialist philosopher Jean-Paul Sartre describes the odd but familiar behavior of a café waiter whom we will call Pierre:

> His movement is quick and forward, a little too precise, a little too rapid. He comes toward the patrons with a step a little too quick. He bends forward a little too eagerly; his voice, his eyes express an interest a little too solicitous for the order of the customer. Finally there he returns, trying to imitate in his walk the inflexible stiffness of some kind of automaton...He applies himself to chaining his movements as if they were mechanisms, the one regulating the other.[29]

Sartre asks why the waiter acts this way, and concludes that he is playing a sort of game. What game? The game of *being* a waiter in a café. Pierre has some vision of what a waiter should be, and he is willing himself to be just that. His vision of perfection includes a script for each occasion, a proper response to each state of affairs. Perhaps he is simply trying to do his job well, but Sartre thinks there is a deeper motivation, a shadow of the core of the human experience: Pierre is trying to extinguish his freedom.

Sartre distinguishes two kinds of existence (actually three but we don't need the third for this demonstration): being-in-itself (*l'être-en-soi*) and being-for-itself (*l'être-pour-soi*). Being-in-itself is the kind of existence that a rock or a chair or a tree possesses. Rocks and chairs are static, stable objects which do not seek to be anything other than what they are. They do not *seek* at all; they simply *are*. In contrast, the being-for-itself is a consciousness, an active agent. Because consciousness is always consciousness of something, the being-for-itself *does* seek; it is defined by its quest to fill itself with other things – specifically things to think *about*.

The goal states of a being-for-itself, the objects of its consciousness, are then, Sartre says, just beings-in-themselves. When you think of a rock, the thought-of rock is a stable, static entity; it is obviously not like you, possessing neither a consciousness that seeks anything nor a will to impose upon the world. And no matter what you are thinking, there is a separation between your consciousness and the object of your consciousness, a chasm of nothingness. From this, Sartre derives a fundamental axiom: each being-*for*-itself tries to cross this chasm and become a being-*in*-itself. Consciousness without an object is nothing. Sartre's paradox: in order to be what it is, consciousness must become what it is not.

Of course, this quest cannot succeed. As long as consciousness is conscious, it is conscious of something. Those two things (mind and its object) are not the same, indeed have different types of existence, and so consciousness is isolated. Pierre's *idea* of waiterhood is a being-in-itself (it is just an idea, a static object, not an independent conscious being). By sculpting his own behavior to that shape, the shape of the idea, Pierre is seeking a state of affairs in which he himself, as Pierre, will perfectly embody that idea. But in doing so, Pierre will not be conscious. For every situation, the action of the perfect waiter is predetermined. If Pierre is successful in this quest to be the perfect waiter, his freedom will be superfluous. He will be released from the burden of having to constantly decide what to do with himself. The emptiness at the core of his being will be full.

This would be quite a relief, if it were possible. It would absolve Pierre of the relentless need to make choices, to decide each second what to do. But even if he temporarily succeeds in embodying the perfect waiter, he must choose to renew his decision to continue that effort each moment. With every arriving second, Pierre must pretend to sublimate his freedom and instead fill himself with the idea of perfect waiterhood. That pretension is in itself a decision, so according to Sartre, it is willful blindness to his true nature. Pierre is living in *mauvaise foi, i.e.* bad faith.

Although most of us may not try to embody so singular an idea as waiterhood for such long periods as Pierre, we take essentially the same pill in smaller doses. At the office, we hope for tasks with clear processes and definitive end states. Making decisions against unspecified success criteria is stressful. As the host of a gathering, we hope to appear calm and collected (our idea of perfect host-hood). A longer-term, perhaps quasi-religious belief in *destiny* is really the desire for release from the pressure of thinking and acting constantly to determine one's own fate. In the ultimate extension, we are "haunted by a vision of 'completion'" which religions call God and which Sartre calls *ens causa sui* ("being one's own cause").[30] By being simultaneously a mind and a motionless thing (motionless because there is no separation between Him and His creation), God is what a human being wishes to become. When He perceives something, God perceives merely a part of himself. Men, in contrast, are always aware uniquely of what they are not, always separate from the other, in motion toward it but never reaching it.

One needn't accept Sartre's metaphysics to understand the first of these quests, *i.e.* why indecision causes anxiety. The biological explanation is simply that uncertainty about what to do is dangerous, for the same reasons that failing to understand what you are looking at is dangerous. Because they are unsafe, evolution has made these experiences unpleasant. Thus we must ask: can we understand how and why, from a neurobiological point of view, indecision causes anxiety, and thereby get a neural basis of Sartrian philosophy?[31]

For Pierre, the mental model of the perfect waiter is composed of scripts, *i.e.* algorithms, directing behavior in various circumstances. Ideally, the model has a script for every contingency. Where do these scripts live? Like other motor plans, they live in the frontal lobe – or possibly in a downstream system like the basal ganglia to which the frontal lobe has offloaded the program. In the framework of neurobiology, Pierre's bad faith is just the desire to have a preset action plan that he knows will be appropriate in any circumstance. Without such a plan, his frontal cortex must do more computation, forecasting different possible behaviors to try and predict future payoffs. While such computations are being done, Pierre is in a state of indecision, and this is dangerous and therefore stressful. He wishes to avoid this state in the future. This is the neurobiological origin of bad faith.

Sartrian *mauvaise foi* can also be applied to subsets of the human experience, such as art. The goal of the artist is to capture a piece of his or her soul in the work. Existentially, this is done so that when someone else experiences and understands the work, the artist's soul will live on in that viewer. It is a kind of spiritual procreation, or memetic parasitism, beginning in the mind of the artist, incubating in the creation of the work, and then maturing in the mind of the viewer. Through this process, the artist hopes to achieve immortality. But something fundamental is lost in the transmission: the artist's subjectivity. In software lingo, this is a call by value, not by reference. Some shadow of the spirit lives on, but it is a copy only; the artist him- or herself does not have the experience of actually existing as an active agent in the work or the minds of the viewers. Like Pierre, the artist is engaged in a Sartrian quest, with the mind as *l'être-pour-soi* and the work as *l'être-en-soi*. The act of creation is a yearning across the chasm of nothingness, and the illusion that the chasm has been crossed is *mauvaise foi*. How can we understand in neural terms *this* Sartrian project?

Given any state of mind of an artist, many means of expression are available, any number of possible incarnations of that state. Michelangelo's David did not have to be in that particular pose, or with those particular hands, and we cannot but doubt that Michelangelo gave consideration to which portrayal would best achieve the desired effect. The artist must choose among all the possible works that might represent a given state of mind and select that unique one which does the job the best. If this is achieved, the artist will feel that the work is the most beautiful expression of the associated idea. But what does this mean, and how does beauty relate to the Sartrian quest? Just here: the artist starts with an idea and attempts to back-project the idea onto some sensory domain – it could be auditory, as in the case of music, visual for painting, spatial for sculpture, *etc.* He or she does this again and again, struggling to find that hypostatized stimulus that will produce most purely the idea. The neurologist Zeki makes the point in this way:

> Plato implied that, at least to get nearer to the Ideal, painting should change direction in order to represent as many facets of an object or situation as possible, since this would give more knowledge about the object. What Plato only implied, Schopenhauer made explicit many centuries later when he wrote that painting should strive 'to

obtain knowledge of an object, not as a particular thing but as Platonic Ideal, that is, the enduring form of this whole species of things', a statement that a modern neurobiologist could easily accommodate...

Implicitly more dependent upon brain function [however], and thus more acceptable neurobiologically, are the views of Kant and Hegel. ...The [Hegelian] Idea, then, is merely the external representation of the Concept that is in the brain, the Concept that it has derived from ephemeral sense data. It is, in fact, the product of the artist. Art, including painting, therefore, 'furnishes us with the things themselves, but out of the inner life of the mind.'[32]

But this selection of the strongest connection of a stimulus with the neural state representing an Idea is exactly what we have described elsewhere, as in the meaning of a word. The meaning of a word is the center of mass of the associated perceptual cloud, *i.e.* a prototype. This is why it is natural to ask what a given work of art means, even though one would not naturally ask the meaning of the *referent* of the painting, *i.e.* the object or person or scene itself. According to the art critic Arthur Danto, this is the very essence of art: "*x* is an art work if it embodies a meaning" (*The Abuse of Beauty*, p. 25).

For many people, meaning is the scientific worldview's Achilles heel; it is precisely what science is not supposed to be able to provide. But meaning consists in the linkage between a mental symbol and its referent, such as a percept or concept. When such a percept or concept is prototypical of the collection of possible corresponding things (this should be a familiar refrain by now), strong binding and rhythmicity are achieved quickly, and reward centers are activated. Thus the creation of a work of art is simultaneously a search for beauty, an expression of meaning, a doomed attempt to complete a

Sartrian quest, a realization of the Hegelian Idea, and self-medication with serotonin.

Kierkegaard

Writing in the mid-1800s, the Danish philosopher and theologian Søren Kierkegaard preceded Sartre by over a hundred years. Despite his devoutly Christian – and Sartre's atheist – perspective, he created a foundation that Sartre would use, including notions of the centrality of despair which would lead to Sartre's conceptualization and invocation of bad faith. The perspective on Kierkegaard given here is less neural than those for other philosophers, and rather more of a suggestion of how some of his more beautiful and melancholic contemplations can still be entertained under a naturalistic worldview.

Kierkegaard's 1843 book *Fear and Trembling* is an exploration of the faith dilemma faced by Abraham when an angel commands him to sacrifice his son Isaac (Genesis 22). On one hand, Abraham's acceptance of the command, and his willingness to perform the sacrifice, anticipates the costs of faith presented in the next chapter. On the other hand, faith that comes without paying a price is, for Kierkegaard, hollow. It is the abandonment of reason that Abraham must accept, and which Kierkegaard accepts via his willingness to believe the absurd and act on it. Only by believing or doing something *ex nihilo*, something utterly lacking in logical merit, can either man demonstrate to himself and to God of the strength of his faith.

Faith, in other words, requires suffering. It's not that Kierkegaard has found a way to be sure of his religious beliefs; it is precisely the fact that he is *not* sure that gives his faith its vitality. It was that very notion, the placement of suffering at the center of the human condition (both a very Christian and very Buddhist idea, though for different reasons), which proved so influential to Sartre. For Kierkegaard, suffering comes from faith. For Sartre, it comes from *choice*. The naturalist, oddly, may be closer to Kierkegaard, seeing that (a) the libertarian freedom that Sartre posits is an illusion, and (b) the yearning for beauty and order, and the hope that we will find it, is reminiscent of the longing for God. Under the framework presented here, our brains extract ideals from experience, then search for them in future experience, hoping for but rarely finding that aesthetic ideal.

With the reader's indulgence and a little interpretive latitude, we might also profitably become neurological literary critics. As Kierkegaardian metaphors and products of the aesthetic quest, consider H.G. Wells' *The Island of Dr. Moreau* and Mary Shelley's *Frankenstein*. At first glance, these are the stories of mad scientists and their monsters, but they can also be seen as allegories of the human condition and pain (Sartre's despair, Kierkegaard's anguish) as its essence. In each novel, the pivotal character is a creator god and is responsible for the pain involved in the formation of human life. Moreau is straight out of the Old Testament. Feared, revered, preeminently knowledgeable, he is described as a "massive, white-haired man," "powerfully built," who knowingly causes horrible suffering in his children: The seedy, unsterilized hut where the operations are performed is the "House of Pain." In *Frankenstein*, Victor intends his creation to be beautiful but fails terribly, dooming the creature to a lifetime of isolation and shame. Secondly, each book highlights the psychic pain of being not quite human. Each of Moreau's creations struggles with its animal nature and suffers trying to overcome it, to obey the Law, until at last it dies, its dreams unrealized. As the only one of its kind, Frankenstein's creature is half man, painfully aware of his separation from the rest of humanity and desiring completion in a female companion – a request ultimately denied by his creator. Just as Frankenstein's fiend commits murder, and as Moreau's creations ultimately kill their creator out of their frustrations with imperfection, so we humans kill God, as Jesus.[33]

Do Moreau or Victor Frankenstein sin in their acts of creation? Moreau knows that his children will suffer not just in the act of creation, but also *simply by existing*. All he need do to prevent this pain is discontinue the experiments, but he does not, and we despise him for it. It is the same with Frankenstein – we are supposed to be appalled at his presumption and at his impulsive and visceral rejection of his creation, his total lack of responsibility. If there is any substance to these metaphors, a Kierkegaardian philosophy would seem to extend these question to that of whether God sinned when he created man. If so, did he atone for that sin by offering his son, his only son Jesus, whom he loved, as a sacrifice? But then, knowing that His son would suffer, did God sin in sending him here to atone for ours? For Wells and Shelley and Christianity at least, there seems to be no way out.

Whether interpreted as philosophy or allegorical fiction, these works cite human suffering as a center of gravity around which moral philosophy is constructed. If naturalism seeks to replace religion or metaphysics as point of reference, or if indeed it hopes to be a lens through which all human activity can be seen, it must offer an explanation of the source of these works. That explanation is this: suffering as incomplete and imperfect beings, we seek God; that is, the beauty and unity of all things. But it is that unity, the universe, that has produced us in the first place and condemned us to this quest. The universe gives birth to its children who will necessarily suffer, yet as a non-conscious *être-en-soi* itself, there is no way it can know that it is causing pain. Minds are the vehicles of discovery; without them there is no knowledge, but with them there must be suffering. So God imposes pain but does so unknowingly. Only through the suffering of human beings can the universe come to know itself and see what it has done: only through Christ's pain can God discover himself.

"For if the immediate and direct purpose of life is not suffering, then life is the most ill-adapted to its purpose in the world."

– A. Schopenhauer

"The dual substance of Christ – the yearning, so human, so superhuman, of man to attain to God or, more exactly, to return to God and identify himself with him – has always been a deep inscrutable mystery to me. This nostalgia for God, at once so mysterious and so real, has opened in me large wounds and also large flowing springs.

My principal anguish and the source of all my joys and sorrows from my youth onward has been the incessant, merciless battle between the spirit and the flesh. Within me are the dark immemorial forces of the Evil One, human and pre-human; within me too are the luminous forces, human and pre-human, of God — and my soul is the arena where these two armies have clashed and met."

– N. Kazantzakis, *The Last Temptation of Christ*

What can we accomplish by building such connections? First, we can try to explain the natural origin of this literature – philosophy both pure and disguised as fiction. Only if such projects succeed can we justifiably claim that naturalism *could* contain a foundation for all the thought-products of humanity. Secondly, this work is beautiful, both as literature and idea. If, when interpreted through the naturalistic lens, this literature can still reveal aesthetic philosophical treasures, so much the better – it shows again that a scientifically grounded outlook can possess the same warmth and poetry as any of its competitors. Some of the authors in focus here (Kierkegaard at least) would likely reject the naturalistic interpretation, but perhaps not the motivation for such an effort.

Plato and Aristotle

Plato believed that the ultimately real was a realm of ideal forms or Ideas, of which the ordinary objects that we see are merely imperfect and confused manifestations. It was a realm of purity, beauty, and the divine (Plato tellingly uses these words many times in the *Phaedo*), logically prior to the world we know. The ultimate reality is the realm of the gods and the place where the soul would dwell after death, *i.e.* heaven.[34] Despite important disagreements, Plato's student Aristotle maintained the existence of the pure forms, calling them "universals," but believed that they were extracted from the world we see, not existing independently or giving rise to it.

The difference between these two philosophies is so close to the difference between those of the religionist and naturalist – at least, as their views have been presented here – that we must feel inclined to agree with Whitehead when he said that "the safest general characterization of the European philosophical tradition is that it consists of a series of footnotes to Plato." The notion of perfection as having an external reality is Platonic and most amenable to the supernaturalist; the notion of perfection as existing only in our minds, which extract the Idea from experience, is Aristotelian and agreeable to the naturalist. Much of what I have said here is a neural and cognitive recasting of these two philosophies, but this is not to say that that recasting is unimportant – this for reasons already given.

To see this mapping more clearly, consider once more how we learn nouns. In early experience, we begin seeing things – apples, for instance. Repeatedly our attention is directed to an object and, in close association with it, to the word "apple," spoken or written. Because each actual apple is slightly different, each trial excites slightly different neural activity patterns, and we can imagine the mental representation of the collection of apples as a cloud in neural activation space. The same is true for the word stimuli, although in that case the variability is smaller. By virtue of the two sets of neurons – one representing the signified (the apple), and one representing the signifier (the word) – the synapses between the two populations are strengthened. After training, activation of one will excite the other.

Now let us ask: What does the word "apple" *mean*? It is probably important to understand what we mean by saying that something *means* something, but set that aside for a moment. Just take as a provisional definition the question, "What kind of object do we imagine when someone says 'apple'?" Because of the association we've learned, the sound or appearance of the word excites various sensory representations in the cloud to different degrees. We probably have seen some apples slightly distorted or discolored, or otherwise unusual, and the word is not going to call these to mind to a very great extent. We will also have seen relatively more often apples of no great distortion or discoloration but more or less typical. Does the word excite activation of any of these to the exclusion of the others? Probably not. More likely, what happens is that all feature detectors are activated in proportion to their frequency of occurrence during training, and the percept corresponding to the word, *i.e.* what we imagine when we encounter the word, is a weighted average, a prototypical apple, the percept represented by center of mass of the cloud of apple representations.

Disregard for a moment the fact that language can be used to communicate and focus on its use as an internal model-building tool. Even if you think in some unspeakable "mentalese" rather than your native language, the mentalese token for this fruit will still point to the prototype. This is how both language and cognition come to rest on a foundation of ideal or Platonic forms – and because of the connections between prototypes, binding, and reward, it is not surprising that Plato felt that this world of the Idea was beautiful. When we question ourselves on the essential nature of the world, we will be tempted to think that the prototypical apple is somehow the underlying reality to which real objects are approximations. Though we may never actually encounter the perfect apple, it seems to inhere in our experience. The same goes for prototypical faces and prototypical chairs; there seems to be a perfection hiding behind the mundane. As it turns out, Aristotle was on the truer path. There is no such heaven in reality any more than there is whisky in wheat: it's only there because we distill it.

Wittgenstein

> Most of the propositions and questions of philosophers arise from our failure to understand the logic of our language. (They belong to the same class as the question whether the good is more or less identical than the beautiful.) And it is not surprising that the deepest problems are in fact not problems at all.
>
> – L. Wittgenstein, *Tractatus*, 4.003

Now comes an important question. Do apples exist? On the face of it, this is a ridiculous question. Of course apples exist! *Something* out there is responsible for my experience of apples. What is that something? Apples, of course.

This makes perfect sense when we use language in the conventional way. Conventional use is rough and ready, eminently practical, and sufficient for its original purpose. Wittgenstein, however, went beyond the everyday and examined language, in particular the question of *meaning* we mentioned above, more analytically. Because he was then using language to talk *about* language, and sometimes finding fault with it, readers can find this process somewhat exasperating, akin to sawing the branch one is sitting on. Things will get even a little bit weirder here because I am interested in using a neural / cognitive toolkit to analyze why

Wittgenstein is deriving the philosophy that analyzes language. But let's just see what happens.

Do apples exist? As pragmatic realists, we can at least agree that we saw *something* out there; we didn't hallucinate all the experiences upon which our acquiring language is based. But are the things we saw apples? If it weren't for our sitting around labeling things, would the universe know what those objects were? It seems safe enough to say that the universe would not go to the trouble to actually *label* them. After all, what language would it speak? Would the universe think of them as apples or manzanas? Would there be invisible little stickers on everything, labels with every possible word that someone *might* use to denote them? Is the universe a linguistic, thinking entity? No. Clearly the universe doesn't think of them in words...but perhaps that's not what we mean when we say "apples exist." Perhaps we can rescue the question.

Perhaps what we mean is that all the things which we call apples are related to each other in some way, and while the universe doesn't go to the trouble to think about them linguistically, it treats them all alike. *I.e.* perhaps applehood resides in facts shared by all the would-be apples. We might reconcile that idea with a naturalist foundation by saying that it is in virtue of shared physical characteristics that these objects behave in certain predictable ways. They have similar masses, molecular compositions and so forth, and in that sense the universe "knows" what to do with them.

Mass? Molecular composition? Perhaps we should convince ourselves that when we're holding what we think of as an apple, we know exactly what we have. We suppose that it is spatially circumscribed, for example, but in so doing we ignore evaporation, oxidation, thermodynamic movements, *etc.* We know that other apples would not be identical; each would vary in mass, shape, color, sugar content, and so on. Atomic-scale physics cause quite a smear at the borders of an object. While our coarse senses don't perceive such effects, the universe does, and acts accordingly. The universe doesn't even "think of" what we're holding as an object *per se*, let alone an apple. It simply acts on individual particles. In the classic *Gedankenexperiment*, we could remove one atom at a time from our apple, eventually reaching a point where we held a single molecule of fructose. Was there a unique moment at which the thing we held stopped being an apple? This prospect alone should prompt us to question the likelihood of constructing a physically sensible theory of meaning – at least one that includes the naïve idea that meaning consists in pointing to physical objects. Indeed, it is physical variation – albeit of a coarser variety – that presumably motivated us to consolidate the many individual fruits into one group and give it a single linguistic token. Rather than storing and communicating all the details of each particular apple, one can simply say "I have an apple," and a number of features, accurate to within some range, can be inferred. We must understand, however, that the universe does not make these simplifications.

Nevertheless, we can verify through experimentation that the universe does indeed treat more or less similarly all those things we call apples. But still: is this the same thing as saying that apples exist?

In a roundabout way, we have arrived at a sense of the question often asked by philosophers of religion. If we think of something, does that reify it? Does God exist in virtue of the plain fact that some people believe in Him? Until this point, our naturalistic requirement for existence is that the thing in question have at least one physical occurrence. "Do apples exist?" is canonically affirmed, in contrast to the question, "Do unicorns exist?" Since unicorns are not really "out there," we typically say no. But as we've seen, it appears nontrivial to even say that *apples* exist under a strict physicalist interpretation. The objects that produce the perceptions to which we refer when we say "apple" are out there in the world, but they are not united under any banners of applehood. They have no substantive allegiance to each other; the fact that they have similar origins and compositions and behaviors is merely a happy accident. In this light we might assert the existence of the objects that produce apple-like perceptions but refuse to call them "apples." *The things that we call apples* exist, but apples don't.

Returning to theology, we can see a second interpretation of an existential claim. Does the thing in question have a referent in at least one mind? For example, "Does happiness exist?" might be answered affirmatively if we knew that some conscious being felt, according to itself, happy. Of course happiness exists, at least in some sense, since I am thinking about it *right now*. Unicorns exist as well, just not out there. Unfortunately this interpretation too has a problem. The referent whose existence is being probed comes from the mind of the questioner. Can someone else possess my happiness? It would seem not. Or suppose that, in a state of unhappiness, I ask, "Does happiness exist?" Since I am not happy, I must be asking whether happiness as I conceive it exists anywhere else. (I certainly am not asking whether happiness as *you* conceive it exists...I can't even know how you conceive of it).

Now, relative to atoms, neurons are large objects. Their membranes are fluid-like and highly dynamic, their electrochemistry constantly in flux, and even their DNA is changing. We can essentially guarantee that no single neuron will ever be in precisely the same state twice; this is even more true with an ensemble of many neurons. Since it is those neurons that support the cognitive representation of the word "apple," no two cognitive representations of the word "apple" can be the same either. So how could my brain state, which can't even be replicated twice in my own brain, be duplicated in an entirely different brain? On a physicalist interpretation this seems impossible. The only way to duplicate a state across two people would be to forget about the physical and resort to some sort of abstract model, but that would beg the question. Our answer must be no. No matter who asks the question, happiness never exists anywhere else.

On the other hand, a questioner must at least partially activate the brain state of happiness when posing the question. Just as you cannot suppress the concept of an elephant when thinking of the word "elephant," you conjure some neural support for happiness just by asking yourself whether it exists. In the moment that you ask the question, happiness does exist, for you, at that moment. But the next time you ask the question, the neural representation for the word will be slightly different – there are too many degrees of freedom to *precisely* duplicate a neural state – so the conjured emotion will be slightly different. Happiness will exist again, but it will not be precisely the same happiness as before.

Then there are questions concerning the existence of things that seem to reside somewhere between the outside world and our minds, like numbers. Does four exist? This is easy to answer if we remind ourselves of our physicalist foundations. The word I utter when I ask the question has a referent in that moment. So in that moment, four necessarily exists. The next time I ask the question, the referent will be slightly different, and will exist again, but it will not be the same four as before. Although one might be inclined to say that the properties of the number four are much crisper than those of an apple or a unicorn, the neural representation of those properties is, for all practical purposes, equally fuzzy. If the thought is instantiated by neurons, it is not precisely replicable. To be sure, there are many foursomes of things out there in the universe, but that is not the same thing as saying that four, four *itself*, exists.

So far, the only really solid affirmation of existence we've seen is that it is guaranteed – indeed it is brought about – merely by the asking of the question. Such existence is evanescent, lasting only as long as the neural trace, but at least it's there. But in that sense, the existence of an intentional object is a tautology, and as such existential predication adds nothing. We may as well remove it entirely and probe the meaning of the question "Does four?" Or even more simply, "Four?" Although ungrammatical in English and probably mentalese as well, this is philosophically intelligible. In some ways it is even cleaner and clearer that the original. Four? Yes. – But not, "yes, four," because the repetition is imperfect. Seen in this way, the entire idea of existence disappears from philosophically meaningful dialog. Existence is the one predicate that has no meaning.[35] Here, of all people, is William James:

> Higher stages still of contemplation are mentioned – a region where there exists nothing, and where the [meditator] says: "There exists absolutely nothing," and stops. Then he reaches another region where he says: "There are neither ideas nor absence of ideas," and stops again. Then another region where, "having reached the end of both idea and perception, he stops finally." This would seem to be, not yet Nirvana, but as close an approach to it as this life affords. (pp. 401-2)

This description of existentially null phenomenology closes the loop with the neural state described (or posited) in Chapter 1 as the counterpoint to the unifying religious experience.

We can now ask another Wittgensteinian question: what is the nature of the relationship between the word "apple" and a particular apple? In other words, what is meaning?

Consider the glyph "apple," *i.e.* its visual manifestation on the screen or page where you are now reading it. In the electronic domain, each phosphor of the screen contains millions of atoms. From phosphor to phosphor, those atoms have slightly different relationships with each other. Many millions of photons stream through each pixel, or reflect off each patch of paper, every millisecond. Thus no two pixels are identical, let alone two characters or words on the screen. Likewise on paper. The ink laid down by a pen does not create Euclidian lines. Small variations in ink flow, bonding of the ink with the paper, orientation of the fibers *etc.* mean that each glyph is physically unique. And again, no two verbal utterances of "apple" are identical. Variations in airflow produce a large and non-repeating state space. No two verbalizations will produce identical states of affairs.

We are left with the conclusion that each particular occurrence of the word "apple" is not in fact identical to the occurrences that have come before, but rather, manifests itself physically uniquely right here on Earth, as does each apple. Each instance of the word might be a sound, an arrangement of ink particles or a pattern of photons, but it is always a physical phenomenon. Since no two manifestations of the word are identical, the symbol "apple", which was supposed to simplify things by labeling with a singleton a large set of "messy" physical objects is, itself, just another large set of messy physical objects. So the labeling of a thing (or a class) with a word does not necessarily achieve anything in information theoretic terms. Symbols are of the same epistemological status and existential class as objects, and at the end of the day we may achieve no real compression by associating one object with another.

A resilient naturalist philosopher of language may still wish to build an association between these two sets. In that case, we must ask what we mean by "associating" two sets of physical objects. In a physicalist worldview, the association itself would need to find support in natural phenomena. Since the *origin* of the association is in our minds (we invent language), not on some ethereal linguistic plane, the *conventional* understanding of meaning would seem to require the positing of some form of telekinesis capable of instantaneously creating physical links between infinite collections of spatially segregated objects. This is patently ridiculous. As long as we remain "outside the head," a physicalist interpretation of reference is not in the offing. And yet, cognitively, it feels possible and quite ordinary to use "a symbol" to denote "a class of objects." Where to next? Obviously, we must move inside the head.

Non-repeating "word-like" stimuli produce one set of non-repeating neural representations. Non-repeating "object-like" stimuli produce a different set of non-repeating neural representations. Each time two elements of these respective sets are associated, *i.e.* each time two populations of neurons are activated stimultaneously, the synapses between the component neurons become strengthened. Reference, a.k.a. association, is thus an observable physical phenomenon, supported with anatomical changes, the addition of mitochondria at the presynaptic terminal, increased production of neurotransmitter, increased postsynaptic receptor production, and so on. Because most neurons within one population will be activated to some extent by stimuli that excite the other neurons in that same set, and because phenomenological linguistic experience is of the center of mass (the prototype) of these two populations, our macroscopic sense that "apple" means apple emerges as a statistical effect.

In summary, the process by which "apple"-like stimuli bring to mind apple-like percepts has been given a physical basis. In exchange, we have had to abandon the independent reality of pure symbols, pure objects, and reference and meaning *per se*. All we have left are physical states of affairs, and specifically, for any discussion of language, meaning, and reference, states of the brain. I say "pure" objects to make clear two closely related points. (1) The objects producing apple-like percepts exist, of course, but they do not participate in reference relations; and (2) as such, the objects have no link to the naming identity that we give them, including their participation in a class. Thus they have no relationship to each other. There is no such thing as an apple, except in our mind. In the world, there are only individuals, each unique and unlabeled. A corollary is that objects are not what we are talking about when we talk. This is what we wish to do, but it is unfortunately beyond our ability. All we are actually talking about when we talk are perceptions, or more generally, states of mind.

If you are impatient with such philosophizing, remember that although non-philosophers don't commonly encounter silly armchair questions like "Do apples exist?" such ordinary civilians *do* encounter questions like "When does life begin?" or "Do animals have rights?" or "Is this man competent to stand trial?" Understanding what we mean by meaning has powerful practical consequences. Whatever our answers to these particular questions, the decision process in general is carried out by (1) reflecting upon the word or idea in question and (2) projecting that word back onto the sensory or conceptual representation it signifies. This process deals only with our own private experiences and thoughts, not about something that exists in the world. Questions like "what is justice in this case?" or "is murder evil?" have nothing to do with world itself; they deal only with our experience, our private (or shared, as the case may be) values, and the way we use words.

*   *   *

Philosophical transparency was apparently not a high priority for evolutionary adaptation, but entropy reduction was, especially if a nervous system was to be of any use in interacting with the world. In fact, entropy reduction is almost inevitable, even in the simplest creature. Imagine an animal whose entire sensory system is composed of a single photodetector. Suppose this photodetector has a pigment tuned to a particular electromagnetic frequency. What is the function describing the cell's response to other frequencies? Macroscopic natural phenomena like this are almost never discrete, so the cell's response is not a delta function (a sharp spike of activity for the preferred stimulus but no activity at all for any other frequency), but instead is at least continuous, probably something like a Gaussian, or bell curve. Now suppose in a certain moment we see the sensor activity is a particular value. What is the corresponding world state? We cannot say with certainty. There is undoubtedly some noise, there is firing rate accommodation, and there are other neural mechanisms that make it impossible to say for sure what the state of the world is. Thus, this single detector is responsible for the creation of some equivalence class: it lumps together a range of stimuli into a single value. It is a many-to-one mapping, which implies a loss in information, a reduction in the entropy of sensory experience. If it was a one-to-one mapping, entropy would not change, but then we might as well not have a sensory system. It is like Stephen Wright's joke about a map of the United States where the scale is 1:1, where one mile equals one mile. A perfectly accurate map is of no use. Some information must be discarded.

What is discarded is the content that makes each individual unique. We always try to move past uniqueness, classifying, trying to see each particular occurrence as an instance of some ideal form: this is Plato. We do this not only for noun-like things (apples, chairs, dogs), which we attempt to place into static categorical theories, but also for verb-like things (trajectories, behaviors of people, defensive strategies), which we attempt to place into dynamic theories. All of life is an attempt to make experience predictable: this is Sartre. We gather information, collecting and sorting it into bins so that one day, everything will be predictable and we will not need to collect or sort any more. We do this not out of some metaphysical quest, but because it confers an evolutionary advantage. Animals that can predict the world fare better than those that cannot.

There is a universe, but its properties cannot be investigated with language. That is to say, God exists, but he is inaccessible.[36]

> The tao that can be described
> is not the eternal Tao.
> The name that can be spoken
> is not the eternal Name.
>   – *Tao Te Ching*

Russell

There is a famous logical problem, often referred to as "the Barber paradox," that goes like this. In a certain town where all men are clean shaven, there exists a barber who shaves all those *and only those* who do not shave themselves. In other words, if we take any man from this town and find that he does not shave himself, then the barber in question shaves him. Also if we find any man who *does* shave himself, then this barber does not shave him. Sounds fine, right? The problem is, who shaves the barber?

If the barber shaves himself, then due to the special restrictions of his clientele (he shaves only those who do not shave themselves), he cannot shave himself. This is a contradiction, so we must conclude that he does not shave himself. But in that case, he qualifies as one of those men who does not shave himself, so again to satisfy the definition of his clients (he shaves *all* those who don't shave themselves), he must shave himself. This again is a contradiction. Thus this barber neither can nor cannot shave himself.

The problem was formulated in this way by the mathematician / logician Bertrand Russell. The problem is really an English rephrasing of a problem called more generally "the Russell paradox," which can be thought of in the following terms. Let S be the set of all things that do not contain themselves as members. Does S contain itself? If yes, then S contains itself, so by virtue of its defining predicate, it must not contain itself. If no, then S does not contain itself, so again by virtue of its definition it must. As with the barber, we have a contradiction either way.

The origin of this problem is well understood as being a consequence of one of the axioms of what is called, undisparagingly, "naïve set theory." The axiom in question says that for whatever attribute (*i.e.* predicate) we can think of, there exists a set composed of all those things satisfying that predicate. This sounds straightforward for simple predicates like "is red" or "is polysyllabic" but, as we've seen, things get tricky with more exotic predicates, including "does not contain itself as a member."

Several solutions to the problem have been proposed, including one by Russell himself, but the most popular is from Ernst Zermelo and Abraham Fraenkel. Despite its formidable name ("the axiom schema of separation") it is fairly straightforward. It denies that you can just make up a predicate and expect there to be a corresponding set. Instead, it says that all you can do is make new sets out of old ones by restricting the membership of the new set to those elements of the old set that satisfy your predicate. In essence, Russell's paradox arises from the attempt to construct sets from thin air. The Zermelo Fraenkel solution is to forbid such conjuring activities and allow only the paring down of pre-existing sets.

What I would like to do here is show how at least some of the motivations for the Zermelo Fraenkel solution are motivated by neurobiology and cognitive machinery, rather than by a disembodied and purely logical quest.

From a physicalist perspective, the outside world is composed only of objects. Predicates (more easily thought of as attributes in this context) must be inferred from the overlapping brain representations of sets of objects. We might get the idea of roundness, for example, from the sensory overlap between a ball, an orange, and the sun. At the same time that we are extracting this attribute, language acquisition encourages us to associate it with the word and concept "round." As a byproduct of this process, we begin to reify attributes, *i.e.* to conceive of the world as being composed of two types of things: objects and attributes.

The cognitive unpacking of the question "Is this object round?" is an investigation of whether this object excites the same neural ensemble as did the foundational objects from which roundness was extracted. While this makes sense for features that actually were extracted from experience, problems arise for predicates constructed *ex nihilo*. How, for instance, are we to interpret predicates like "is married to a tricycle-riding anteater"? This predicate is surely not derived from experience with objects that satisfy it, so what is the status of its corresponding set? Does it exist, and simply have no members? Or does it not exist at all?

When a mind built on these foundations considers the predicate "does not contain itself as a member," the inclination is to reify it, as we do with colors, shapes, and other attributes. Once reified, it seems a simple enough matter to collect all the "objects" (in this case, sets) that satisfy it. But this is dangerous because, per the above, the predicate's only real reference is in our heads, amongst the objects for which it originally obtained. Since Russell's hypostatized collection of sets satisfying the function was not originally part of the function's abstraction, we might therefore refuse to continue on grounds of meaninglessness.

All of this might seem to preclude the existence of anything that we do not already know to be true. Strange as that may sound, it is very reminiscent of the first three major propositions of Wittgenstein's *Tractatus* (*e.g.* Proposition 1, "The world is all that is the case") – and it is no coincidence that these propositions build up to and explain the meta-linguistic origins of the Russell paradox (Proposition 3.333), nor is it a coincidence that Russell and Wittgenstein were mutually influential collaborators. Overall it would seem that our ability to imagine references for things constructed from the whole cloth of language is the source of a great capacity for speculation and theorizing, but also of a confusion over what constitutes the real.[37]

Zen

There is a story told about Buddha. He had established himself as a sage and attracted a sizeable following. One day a crowd had assembled to hear him speak on the True Teaching and the Wondrous Mind of Nirvana. On his way to the front of the assembly, he noticed a lotus blossom growing and picked it. When he arrived at the head of the crowd, instead of speaking, he merely held the flower up for everyone to see. Minutes passed. Still he said nothing.

Finally understanding, a disciple named Mahakashyapa smiled. Buddha called Mahakashyapa to the front, gave him the lotus flower, and announced to the assembly, "What can be said I have said to you, and what cannot be said I have given to Mahakashyapa."

Twenty eight generations after Mahakashyapa, his ideological heir Boddhisatva brought the essence of this teaching to China, giving birth to Ch'an Buddhism. The tradition spread to Japan, where it became known as Zen.

*   *   *

We like to talk, especially to convey information. The lesson Buddha was trying to impart did not contain any information. Information is the removal of uncertainty, and as such it is a reduction in the entropy of the world. But such a reduction necessarily simplifies through linguistic classification, and those classes are purely constructs of our minds. If you want to understand the true nature of the world, or of experience, what you precisely must *not* do is simplify the world, or throw away details. God is in the details. To truly see what is before you, you must stop talking. Stop naming, classifying, simplifying, throwing away. Clear your mind and simply accept what lies before you as it is, not as you understand it to be. Theravada Buddhism, including Zen, and the dhana yoga tradition of Hinduism all share the essential idea that in order to achieve enlightenment, one must eliminate the normal division between self and object, mind and world. In Zen, especially Rinzai Zen, adherence to symbolic / linguistic structures is seen as a central obstacle in this quest. Thus koan practice is designed to addle the linguistic mind, encouraging it to shut down and allow the underlying empty mind to escape "word drunkenness." One can hardly ignore the comparison to Wittgenstein.

There is a sense, however, in which this is not possible. At least, it is not possible to truly be human and stop the symbolic and theoretic mental machinery. If what enlightenment really is about is a state of mutual authenticity with the world, then we mustn't pretend to be anything other than what we are – such a pretense would be *mauvaise foi*. After all, it is when we pretend that the world is something other than what it is that we mistakenly believe we have understood it. To be authentically ourselves, we must do what we do, which is to run our symbolic and theoretic machinery as we move about our business. Thus the Zen instruction, "Before enlightenment, chop wood and carry water. After enlightenment, chop wood and carry water." But then, how then does one achieve anything in Zen? How does one make progress in attaining the state of nirvana, of non-self-aware self-awareness? In Zen, there is no progress. One does not become anything that one was not already before undertaking the study. John Keats expressed this irony in the context of another quest:

> Fame, like a wayward girl, will still be coy
> To those who woo her with too slavish knees,
> But makes surrender to some thoughtless boy,
> And dotes the more upon a heart at ease;
> She is a Gipsey, –will not speak to those
> Who have not learnt to be content without her;
> A Jilt, whose ear was never whisper'd close,
> Who thinks they scandal her who talk about her;
> A very Gipsey is she, Nilus-bom,
> Sister-in-law to jealous Potiphar;
> Ye love-sick Bards! repay her scorn for scorn;
> Ye Artists lovelorn! madmen that ye are!
> Make your best bow to her and bid adieu,
> Then, if she likes it, she will follow you.

Is there a difference in the way unenlightened and enlightened think or act? Yes, but it is not minded. Does one lack Buddha nature while the other has it? The tao that can be described is not the eternal Tao.

What can be said can be said clearly, and the rest we must pass over in silence.

– L. Wittgenstein

CHAPTER 5
Methodology Wars

I n the stillness, two men stand atop a stone temple. For the first time this year, Venus, the celestial symbol of divinity and war, is visible before sunrise. The four winds are in balance, and the savior god Quetzalcoatl must renew his self-sacrifice for the continued existence of humanity. The chief priest wears his usual finery, but the other man, playing the role of Quetzalcoatl's avatar, wears a breastplate of hammered gold, inset with onyx and jade. Sprays of iridescent feathers fan from his headdress. He wears snakeskin sandals and a skirt of jaguar leather. Every part of this costume – the number of ornaments, their shapes and materials, the colors of the feathers – has a specific and deliberate meaning.

The calendar used to predict the date of this occasion was developed over the previous thousand years, and is of almost immaculate accuracy. This particular day is the beginning of a 104 year cycle; to assure glory in future battles, Quetzalcoatl's ritual must be completed on this morning. With a crowd of thousands chanting at the base of the pyramid and the hues of the sky slowly lightening, Venus blinks into existence on the horizon. At that moment, assisted by his aides, the chief priest cuts the beating heart from

Quetzalcoatl's chest. As he sears it on the altar fire, the crowd erupts into an ecstasy of salvation and war.

*   *   *

With apologies to Mesoamerican scholars, something very closely resembling this scene could have actually happened. But what, exactly, is it? Clearly it is a religious ritual, but it is something more as well. The participants must believe in some sort of efficacy for the ritual, *i.e.* that there is a cause-and-effect relationship between these actions and the world around them. The belief that "If I do this, then X will happen" necessarily involves logic and empiricism, so perhaps the ritual contains more science-like thinking than it seems at first. And what should we make of the relationships between mythological symbolism and the factual accuracy of the calendar, or the engineering sophistication needed to build the temple? Isn't there something strange about the union of such spiritual and rational enterprises? Many today may find the dense organic symbolism primitive but the mathematics and architecture advanced. To the people who invented them, these enterprises probably seemed like different aspects of the same thing.

We might also wonder why rituals like this would so excite the emotions of the citizens. The costumes, the sacrifice, not to mention the societal and military consequences that follow, require huge expenditures of money, time, and human life. To pay such a price, and to do so repeatedly over many generations, the believers must be getting something in return. It seems likely that that return went beyond mere entertainment value and included some form of emotional and spiritual poignancy.

Eleventh century followers of Quetzalcoatl were just as solemn and certain about their god as we are of Jesus Christ or Allah. In comparing their worldview to those of a modern theist or scientist, there are two key questions. First, if we disregard the feathers and crocodile teeth, is there any deeper methodological difference between their reasons for belief and ours? Is there any fundamental philosophical difference between Quetzalcoatlism and, for example, Christianity? The cultural and temporal distances involved may make this particular religion seem tribal and alien and perhaps even revolting to modern eyes, but the depth of the symbolism and the trouble they took to participate in it should seem quite familiar. Second, if we reject their claims but not those of surviving religions, on what basis do we do so? These questions will form the initial focus of this chapter.

For the moment, we fast forward a thousand years from that Guatemalan morning and talk about its modern-day legacy. For us, far more than for Quetzalcoatl's devotees, science and religion are separated. In the last five chapters, we have tried to understand how to unify at least certain aspects of their *content* – what some might see as an ironic return to "primitive" thinking. As deliberate as that has been, we must at last, and with equal urgency, understand the necessity of separating their *methods*.

Faith as a double-edged sword

Suppose for a moment that you are religious – a Christian, for instance – and that you have a child whom you teach to accept on faith that certain things are true. These would be things for which the evidence is not overwhelming, or even things for which there is no evidence at all. Examples might include not just the particular tenets of your religion (Christ died for our sins, was resurrected three days later, *etc.*) but also norms of social behavior such as humility, charity, and good will. These latter are all desirable attributes, and the close association between faith and upstanding social behavior encourages you that you are doing something good.

Now imagine another parent, also Christian, but unfortunately also doggedly racist. For whatever reason, this person has come to believe that essentially all members of a certain race are genetically, mentally, and morally inferior. It is a safe bet that this person has no good evidence for their racism; they were simply influenced by racist parents and a like-minded social network, and in time, came to interpret their own experiences through that lens. Racism simply became part of their worldview.

This parent has, like you, a child who has been taught the same norms of good social behavior. Honesty, conscientiousness, charity, forgiveness, *etc.* are all virtues because God says they are. How do we know that this is what God says? The Bible. How do we know the Bible is the word of God? The Bible itself says so. The weakness of the circular reasoning here is immaterial because the lessons of morality are so benign. Besides, this is a religious teaching that has nothing to do with evidence or logic, but with faith.

At some point, imagine that this good child has an experience that could be interpreted racially but is not worth much empirically. Perhaps they watch a TV show in which a member of the race in question commits a crime. The child approaches the parent and relays the story of what they have seen. The parent explains by telling the child that people of this race do horrible things all the time. "Look around," the parent says. "These people are responsible for most of the evil in this country. They're lazy, cheating, lying, good-for-nothing." If the child asks probing questions such as "What about Mr. Smith down the street? He's nice," or "What about when people of our race do the same things...does that make them just as bad?" – in this case the irritated parent will probably not interpret the questions as a serious challenge to their racist beliefs, but will instead dissemble, misdirect, or simply chide the child with admonitions not to ask stupid questions. The eventual outcome of this exchange will likely be that the child follows in their parent's footsteps, accepting the racist explanation. This will happen, in part, because the entire framework for believing a claim without evidence or logic is already in place. The child has learned that some claims do not need supporting data...they are simply True, and life will be much easier if no questions are asked. If it seems true, if it feels true, if a belief reinforces your understanding of the world, accept it.

If you find it implausible that a child would receive negative reinforcement or be casually brushed off from an innocent and guileless inquiry, try going to a random church and asking, in the middle of a sermon, whether God can make a stone so heavy He can't lift it. Refuse to sit down until you get an answer that satisfies you as being logically and empirically rigorous. No matter how much inquiry happens within a faith, it is always faith at its core, and there are always specific claims that are meant to be believed in spite of a total absence of evidence. Kierkegaard, for one, wouldn't have it any other way. Refusal to accept that basic method will not be suffered gladly, partially because those who do have faith recognize that there simply is no logical or empirical answer to the complaint. If there were, it would have been removed as an article of faith long before now. Getting stuck on such issues of logic prevents further reflection, and all the vast literature that follows from faith becomes irrelevant. That's too much dissonance for most people, and that's why you won't get a good answer to the "stone so heavy" question.

This story illustrates a core problem with faith as a method for guiding belief. The seed that grows into baseless racism in the child is the same as the one that, in different soil, blooms into humility, good will, and charity. While the end result of faith is, in the latter case, desired, the means by which that result is achieved are unsound. The same foundational lesson that "some things are just True, stop asking questions," can be and probably will be used to justify other things, perhaps quite subtly and perhaps dangerously. The President knows what he is doing. Don't rock the boat, that's just the way things are done here. Once a person learns that critical thinking and empiricism can be set aside when circumstances require it, or even when they would simply be burdensome, the door is open for any claim whatever to be decorated with the flag of uncontestable truth. And once it is so decorated, that claim is safe from all contradiction. Righteousness goes to the person who shouts loudest.

Strictly speaking, undesirable moral consequences are not a sufficient reason to reject the validity or truth of a position. To do so would be to commit the appeal to consequences fallacy, discussed in detail below. Nevertheless, this example is apropos to an examination of faith as a methodology for three reasons. First, religious faith is often viewed as a prerequisite for moral behavior. A 2002 Pew survey found that 61% of those surveyed believed that children raised in a religious tradition were *more* likely to be moral than children raised without religion.[38] The example of racism shows that religiously grounded morality is at best accidental and at worst oxymoronic. Secondly, faith is not a specific claim in itself, but a method, and the proper way to evaluate a method is by looking at whether it is productive or useful for an intended purpose. Faith is productive in the sense that it can be used to support desirable morals, but this is only because its shots scatter so widely that they would hit literally anything. Finally, the moral undesirability of a racist belief is even less important than its lack of factual support, but neither of these can stop faith from endorsing it. In other words, faith as a method is incapable of separating true claims from false ones, and that is not an appeal to emotional consequence but a logical requirement of any method by which we select beliefs.

A ready defense from moderate religions is that faith does not exonerate us from exercising our moral intelligence. When presented with a problem of relating to others, we must not blindly accept anything, even if it is a religious tenet. God gave us a moral sense and He wants us to use it to do good – Francis Collins actually turns this argument around, saying that human morality is the ultimate proof of God.[39] The problem with this argument is that moral claims, in the context of faith-based belief systems, also rest on faith and therefore don't advance the discussion. Saying that a certain claim is true *period* has the same evidentiary structure regardless of whether that claim is about a moral precept, the origin of the planet, or the supposed inferiority of some ethnic group. Why must we not kill others? Because it's a commandment, but also (under the exercise of our filtering moral intelligence) because murder is wrong. But if we ask *why* murder is wrong, we arrive at the same terminus as if we hadn't engaged our moral sense at all but had just accepted the commandment: murder is wrong because...well, it just is. Even if we are unanimous in our condemnation of murder, we have not extricated ourselves from a faith-based metaphysics, which means that we are still vulnerable to – no, not just vulnerable to, but *protective of* – genocidal, racist, ageist, classist and all related sinkholes. Even when faith is used to support an edict that we, today, think of as morally sound, it cannot claim credit for the victory. In such cases, faith is merely a bad reason to be good.

The toxicity of this way of thinking is made eerily worse by the religious sense that shaken faith is a pitiful thing, akin maybe to a bout of depression or an eating disorder. If a believer confronts an event that challenges their dogmatic beliefs, they are usually overwhelmed by a rush of sympathetic and supportive friends, family, and clergy. Faith must be restored and nurtured. Faith is admirable. Those with unshakable faith are the pillars of the community, the most admired, most committed, nearest to God. We must emulate them. We must continue to believe, no matter how frequent or overwhelming the signs that we are wrong. Indeed, the more the world seems to indicate that we are wrong, the more we must insist that we are not. Kierkegaard had it exactly right when he said that a faith that is justified is not faith – the real prize is in believing something unbelievable. Ask questions if you must, but don't do anything when they go unanswered. Ask, but do not doubt. As many polls indicate, having such a baseless and unsupportable moral "foundation" is now essentially a *requirement* for becoming President of the United

States. If a candidate actually based their ethics on reason, they would be unelectable.

The state of the union

Religions of all types depend upon faith. We might even think of faith as a product of religious factories, sent through the distribution channels of missionaries, literature, and grassroots campaigns The sales speak for themselves. And yet, by changing faith's payload from God to racism, or recalling just a few of its historical applications, its essential risk and hollowness become easy to see. In 2005 about 82% of the US population said they believed in God; 18% said they didn't believe in God or weren't sure.[40] What should worry us is not which particular deity that 82% believes in, but the fact they rely on faith to believe in anything at all.

One needn't be a sword-wielding Crusader to make one's faith a matter of public consequence. The gap between the religious and irreligious has played out in recent years with skirmishes over creationism in the classroom, commandments in the courthouse, pro-choice *vs.* pro-life, and of course Islamic fundamentalism. Again, it's not that religion is inherently bad, it's that it depends on faith, and faith is a Trojan horse. Hiding inside the gift of easy moral pedagogy lurks a dangerous principle. And sadly, there is no version of it that selects out just the "good" beliefs…unless it has defined "good" via terms that are also axiomatic, and that doesn't get us anywhere.

We should probably be surprised and grateful, given such a large percentage gap in favor of religion, that the conflict isn't more overt and more violent, at least domestically. Apparently some credit is due to the founding fathers, who managed to construct a constitutional framework in which religious persecution against the minority is kept to a minimum. In 1700 it was Protestants who were in the minority, making the dangerous pilgrimage to America for religious freedom. Today it's the non-religious who are, to a diminishing extent, being protected by the same Constitution. At the moment, no modern-day American Galileos are threatened with torture or house arrest for challenging church doctrine (knock wood), although some might consider too close for comfort the White House's treatment of environmental scientists whose findings conflict with special interests and/or political desires. And of course, private citizens are always capable of the occasional atrocity, such as violence toward doctors and clinics who provide abortions.

Perhaps one of the reasons that the confrontation isn't more impassioned is that many modern Christians are liberal believers in what Daniel Dennett calls "Timeless Benign Force" rather than "Guy in the Sky" religion. In this mode, religionists give what they see as the rampaging bully of scientific empiricism its own playroom and chemistry set. They then define the boundaries of their own beliefs so that they aren't threatened by evolutionary biology, neuroscience, or cosmology. Often, this personal and quiet faith cherry-picks enough New Testament wisdom to encourage tolerance, grace, kindness, and a social conscience. But again, faith does not deserve any real credit for this behavior unless it can be shown that it was the actual cause of the pro-social behavior. As Christopher Hitchens puts it:

> What he's [(the believer)] saying is that if he ceases to believe in Jesus, he's going to instantly become an immoral person. It's a terrible admission to have made! It's an awful insult to human self-respect to say that. And they don't seem to understand that they give themselves over in that way. It's like saying that nothing would stop me from raping you now if I weren't under the supervision of a heavenly dictator. And I have a higher opinion of myself than that.[41]

Moreover, to be fair to the "bully," is that all the accounting we're going to do? Is anyone going to check what happened in that playroom? Something wonderful, other than higher-resolution DVD players, may have been built. Perhaps some of the same benefits collected by Benign Force theists are available under a scientific and rational, rather than neo-Christian and faith-based view.

*   *   *

This discussion of methodology is *not* an attempt to disprove the existence of God. In fact it has nothing whatsoever to do with God as construed in the Christian sense, any more than it does with Quetzalcoatl or Osiris. Instead, it is an examination of what has to happen inside your head if you are going to believe in the supernatural or use religion as a basis for morality. Most of the problems with such beliefs apply to faith in any religion, so theists seeking a sensible worldview should somehow address them. Needless to say, such address does not include sweeping the problems under the rug and saying that faith is merely "a different way of knowing." We don't accept that kind of talk in any other context and there is no reason to accept it here. If I said I had a "different way of knowing" that the Sydney Opera House was actually a sophisticated computer-generated hologram *cum* mind-control experiment, I would be dismissed as a lunatic. Claims require evidence, not just appeals to alternative epistemologies.

With these problems in mind, what I do wish to demonstrate in this and the forthcoming discussion is that a naturalistic worldview avoids not just the problems of faith but also some of the other logical pitfalls of religion. Some may ask why we should bother with such an inquiry, if the Christian view is good enough. The reason we should bother is that on the other side of the ledger, opposite the catalog of good works, there may be hidden costs of faith, even liberal and science-tolerant faith. Never mind fundamentalism – because it crosses over into the empirical, its claims are easily dispatched. The question is: do Benign Force theists have to make any expensive allowances to accommodate their beliefs, and if so, is the burden of those costs really necessary?

Primer in symbolic logic

As we've seen, the key methodological problem with faith is this: if I am willing to accept unquestioningly some proposition, whatever that happens to be, don't I implicitly endorse others in doing the same thing? And if their belief contradicts mine, what can we do?

It's entirely possible, of course, that one of us actually *is* right, and the other wrong. But in general, faith-based beliefs are either (a) uncheckable or (b) held so tenaciously that evidence is ignored. Either way, the insulation against evidence causes faith to present a serious methodological problem: how do we go about finding the truth? This might not matter in the abstract, but in practice, disagreements on matters of faith lead all too easily into violence, so we might be well served by looking for productive methods of truth-finding that don't simply end in "because I said so" stalemates.

One of the central points of faith is its stability in the face of changing circumstances. That can be comforting, but it is almost guaranteed to drift out of touch with advancing knowledge in other areas. This is why those seeking moral direction or scientific understanding from archaic texts have such difficulty with purely modern problems. Genetic manipulation and knowledge of the fossil record didn't exist when the articles of the world's most popular religions were written, and there is no process – other than overextended and subjective interpretations of the Holy text – to handle new issues.

The best way to proceed in this discussion is to learn some formal logic. Unfortunately, the technical notation used in that subject may distract from its implications, so only a non-technical rendition of it is presented here. For those readers who wish to delve more deeply into this subject and look under the hood, as it were, the essential parts of a Logic 101 course are presented in Appendix B. That material requires no mathematical or philosophical prerequisites, just attention and careful study. Most readers will be rewarded for the effort, as it will clarify the precise nature of the methodological problem above, as well as a number of the other things we'll be talking about. Nevertheless, the next few paragraphs summarize the main point.

We asked above what happens if the beliefs of two people contradict each other. A common and moderate response to that question is that nothing at all happens. As civilized adults, we are capable of tolerating dissent, and we understand that on matters of opinion, others may not believe as we do. This is especially true for religious beliefs, and the prevailing view in the United States and Europe at least is that religious freedom is a key component of individual liberty. To say that such tolerance is important is an understatement. In societies that do not embrace it, terrible oppression and violence usually ensues.

Setting aside for a moment the social aspects of this tolerance, let's examine its purely logic aspects. And to defuse some of the tension, let's temporarily desist from using beliefs about religion and consider something emotionally less charged. Say, for instance, that John takes it as a matter of faith that Robin Hood was left-handed, while Mary claims that he wasn't. It may be that the answer to this actually appears somewhere in literature, but let's assume that it doesn't, and therefore that it is an impossible question to answer. This makes the disagreement truly a question of competing faiths, not one of scholarly investigation.

The tolerant thing would be to chalk up the John-Mary disagreement to a personal difference and move on. Even those who do so, however, believe that at the end of the day, one of them is right and the other is wrong. It just doesn't make sense to say that they are both half-right. Or does it? Some forms of logical tolerance are so liberal that they say, in effect, that it is acceptable for both John and Mary to be right at the same time. Rightness or wrongness on matters of faith is private, and only "logical extremists" feel the need to force their beliefs on others.

In preparing to write this book, I spoke with many non-scientists who regard logical absolutism as unnecessarily aggressive – the scientific bully in the playroom. They embrace instead a sort of postmodernism in which truth itself is constructed by the individual. Derrida's deconstruction of meaning and truth has spread far beyond textual analysis to the point where many people seem to think that reality itself is a private matter. To be sure, there is usually unanimous agreement over many aspects of our "private" realities, such as the notion that we are or are not sitting at a table about to eat lunch. But when things get thorny, as when we are going to force either John or Mary (or both) to be wrong about a cherished belief, especially a religious one, then the postmodernist fence-straddling stance emerges. In this case, the way it manifests is in claims that logic itself has a problem. Logic is a human creation and none of us are servants to it. If it starts to act uppity, telling us what we can and cannot believe, then it's time for us to show who the real boss is. If one of 'em has to go, cherished beliefs or rationality, tough luck for rationality.

This is what logicians call the "Relativist fallacy," and it ultimately amounts to the claim that there is no such thing as objective reality. Some philosophers deny that there is a fallacy and take the claim of no reality very seriously. Indeed, epistemology is a fascinating subject, some aspects of which we considered in the previous chapter. For the moment, let us agree to take the pragmatist's view which, frankly, everyone except philosophers knows is true. There is a reality outside ourselves. That reality may include God, or the Tooth Fairy, or the Flying Spaghetti Monster – we won't assume any specifics from the outset. Religion may or may not understand that reality, science may or may not understand it, but we all believe there's some kind of world out there, independent of what we think about it. If you don't believe that, there's really not much point in talking about anything. To commandeer a phrase from Dennett: if you insist that reality is subjective, or that the universe really and truly is constructed every nanosecond out of the ectoplasm of your understanding, then you are thumbing your nose at the whole inquiry. Goodbye, and I hope to see you back again someday.[42]

There are four reasons to make such a big deal about logic and realism. (1) Faith of any kind – not just religious faith – makes logical contradiction inevitable because of the two-party reconciliation problem. (2) Many forms of religious faith posit a deity that is "beyond logic." An omniscient God, for example, must know today what He will do tomorrow, but this is a constraint on His action. Never mind omnipotence...He is not even free to do what he wants tomorrow, since it was preordained by his knowledge of those future acts today. Only a suspension of logical constraint would seem to release God from such a paradox. (3) Depending on how physics ultimately turns out (recall the discussion in Chapter 3), it is possible that empiricism and logic are identical. In other words, it is possible that the universe we live in is the only logically consistent universe. In that case, any belief system that contradicts experimental results is logically impossible. If, alternatively, faith removes its claims from empiricism completely, then by definition its posits have no consequences in the universe. What, then, is being claimed? Why would you believe in something that has no consequences? Would your belief, in fact, have any substance, or would it amount to not believing anything at all? (4) Logical consistency is very important to scientists. It is so important, in fact, that it is really a better differentiator between the two camps (faith and skepticism) than belief in God. If it turned out that there was a logically and empirically consistent way of thinking about God, many scientists could be persuaded to be theists, or at least deists. The real difference between the two approaches is that faith is a willingness, and science an unwillingness to accept logical contradiction. It's not even that the religious worldview seems "wrong" to a skeptic – it's that it doesn't make any sense. To a skeptic, saying, "When I die my soul will rise into heaven" is like saying, "flibberty gonswam ruffa huffa." What does that even mean? If someone can explain it logically, we scientists are all ears.

The appeal to consequences

If you're among the faithful, you might find the presentation so far unfair or unbalanced. For the average believer today – primarily those in Abrahamic traditions – religious participation may feel quite benevolent. Faith supports people, lifts them up, eases sadness, comforts and reassures. It builds social networks, strengthens communities, and encourages people to get outside themselves. Yes, it can do those things. This is not an argument that a faith-based worldview has no positive consequences. For most religions, there are positive social bonds produced by belief, and an optimistic outlook on life may be easier, which does seem to lead to better health. But this is the disentanglement that we must accept: positive effects do not mean that the beliefs that produced them are true, or unique. Even if church were the most reliable tool in the establishment of a conscientious citizenry, we needn't accept that the principles upon which the church was founded are ideal, or even correct. If a Church of Reason became established over the course of several hundred years, and attracted millions of followers, perhaps it too would become an engine of good works. The point is that there are sources other than religious dogma for reassurance, happiness, comfort, social involvement, and so on; finding a basis those things in a naturalistic worldview is one of my main goals with this book.

When someone claims that a proposition is true because it has good consequences (or, conversely, because believing it false would have undesirable consequences), they are committing a fallacy known as an Appeal to Consequences. For example: [43]

- "I don't think that there will be a nuclear war. If I believed that, I wouldn't be able to get up in the morning. I mean, how depressing."
- "That could never happen to me. If I believed it could, I could never sleep soundly at night."
- "God must exist! If God did not exist, then all basis for morality would be lost and the world would be a horrible place!"

Even when the error is not as blatant as in these examples, anticipation of negative consequences can lead to unconscious bias against a claim. Unfortunately, this is not localized to religion and is in fact a very human, very common error. Perhaps these examples will ring a few bells:

- "If my wife is right about the best location for the geraniums, I will have to dig up that whole side of the garden. Ugh! The geraniums would be better left where they are."
- "If Dr. Smith's theory is correct, my last three years of work will have been a waste. There must be *something* wrong with his experimental design..."

Irrespective of whether the person speaking is a Christian, a scientist, or a gardener, this kind of argument is wrong. However – and take a deep breath here – if we set aside the question of correctness and ask instead questions about *utility*, then the consequences of a belief become very relevant. The criticism of faith given at the beginning of this chapter, *i.e.* that it can just as easily be used to support racism as merit badge ethics – is not subject to the appeal to consequences fallacy because faith itself is not a belief; it is a way of believing. The Appeal to Consequences fallacy applies to specific claims, not methodological dispositions whose consequences are the only measure of their strength. In the examples given above, a number of benefits for religion are claimed. By that logic, saying that religion and faithfulness are desirable traits because they produce these benefits is perfectly fine, but the defendant has then taken the witness stand and must endure cross-examination. It is therefore our duty as diligent investigators of world views to ascertain whether there are also any negative consequences of religion and faith – not just manufactured anecdotal ones like racism, but longer-standing historical correlations. If we find any, these will not imply anything about the truth or falseness of the tenets of the religion in question, but they will put into broader context the claims of individual and social benefit. Selective sampling is not allowed.

Needless to say, there is a long history of horrible events whose causes are or were almost entirely religious; without faith (*i.e.* belief without evidence) many of these conflicts may never have happened, or might at least have taken on a less violent form. Examples: Abortion clinic bombings; the American revolution; the Arab / Israeli conflict; the Aum Shinrikyo poisonings; Aztec religious sacrifices; the Branch Davidian conflict in Waco; the Catholic / Protestant conflict in Northern Ireland; the Crusades; the East Timor conflict; the Heaven's Gate cult suicide; the Huguenots and the French Wars of Religion; the Inquisition; the Indian / Pakistani conflict; Islamic fundamentalism including September 11, 2001; the Jonestown mass suicide; the Kosovo / Yugoslavia conflict; the Ku Klux Klan; the Sunni / Shi'ite conflicts in Iraq; the Tamil / Sinhalese conflict in Sri Lanka; the Thirty Years War; and witch trials.

Proponents of faith are often moved at this point to note that Nazism and Stalinism were irreligious movements, and therefore, that two of the most horrific demonstrations of fascist violence in recent history were caused not by faith, but by lack of faith. This is a specious claim, not least because both examples did in fact have strongly faith-based foundations; the Nazis in an assumed racial superiority and devotion to Nordic / Aryan paganism, Stalin in his capitalization of a czarist link with the Russian Orthodox church and demands of unquestioning nationalistic devotion. Hitchens has done more to expose these themes than I could hope to do here.

Secondly, merely removing faith is not what I am advocating. Rather, I would advocate replacing it with critical thinking and the principles of natural context which emerge from the ideas of the last four chapters. Along those lines, can you think of a single example – not even a regional or national conflict, but *a single act of violence* – that ensued from a disagreement over who was a more committed or devout skeptic? Cautious, empirically-grounded skepticism simply does not lead to the kind of maniacal rage produced by faith. (In fact, this can be a problem for skeptics – by seeing both sides of an issue, they tend more toward indecision, even when action is urgently needed. In Yeats' immortal words "The best lack all conviction, while the worst are full of passionate intensity.") To be fair, I acknowledge that some of the lack of examples in this regard *could* be a sampling problem: perhaps there just have never been sufficiently many skeptics around to get into conflicts with each other. Still, in terms of suborning violence it would be hard to do much worse than faith. That alone would seem a sufficient motivation to try global skepticism for the next two thousand years, just as an experiment.

To round out the balance sheet, we should also consider the question of whether faithless people are capable of producing any benefit to society, or if instead they are self-absorbed egoists interested only in greed, lust, and pride. Elizabeth Cady-Stanton, Andrew Carnegie, Clarence Darrow, Thomas Edison, Albert Einstein, Benjamin Franklin, Sigmund Freud, Galileo Galilei, Ernest Hemingway, Alfred Hitchcock, David Hume, Thomas Jefferson, Helen Keller, Abraham

Lincoln, John Lennon, James Madison, H.L. Mencken, John Stuart Mill, Thomas Paine, Bertrand Russell, Carl Sagan, Jean-Paul Sartre, Charles Schultz, George Bernard Shaw, Gloria Steinem, Mark Twain, and Oscar Wilde were all godless and socially redeemed. That even one such person exists shows that religious faith is not necessary for social beneficence; the prior list of conflicts demonstrates that it is not sufficient. Not all those who are faithful are moral, and not all those who are moral are faithful.

Tentative and falsifiable

For anyone interested in the philosophy of science, one of the frustrating habits of infomercial quackery is the relish with which they use the phrase "clinically proven." There really is no such thing as clinical or scientific proof. Nor are there scientific facts, at least not with the same finality as those in religion. The general public seems to believe that science uncovers the certainties of the universe while religion embraces the uncertain. This is a complete reversal of the actual situation and, if it were intentional disinformation on the part of the church (perhaps it is or was, I do not know), would certainly go down as one of the most perversely ingenious marketing campaigns ever. We have already seen how religion makes claims about truth: it demands faith-based belief on the part of its practitioners. Now let's examine briefly what science says about what it knows and does not know.

Suppose we want to know whether a certain drug lowers blood pressure. The appropriate experimental design would be to divide a group of subjects into two groups. Group #1 gets the drug, group #2 gets a placebo. Neither group knows what it's getting, and even the people administering the drug and gathering results are in the dark about what they're expected to see. This is the so-called "double blind" standard.

All the subjects get their pills and go through the regimen, with frequent measurements of their blood pressure. What do we find? Well, we get two sets of data. A set of data from each group shows us how (if at all) their blood pressure changed following administration of the drug. Let's say we find that group #1, the experimental group that got the real drug, showed a blood pressure reduction significantly larger than that of group #2, the placebo group. Has it now been "scientifically proven" that the drug reduces blood pressure? Nope.

First of all, what do we mean by "reduces blood pressure?" In everyone? Or just most people? Even if every one of the subjects in the experimental group (the group that got the real drug) showed reductions in blood pressure, it's possible that we happened to select, totally by chance, a non-representative sample – perhaps, unbeknownst to us, the drug works better for older patients, and if we tested it again with younger patients it wouldn't work.

Also, reduces blood pressure by how much? We must do statistics to decide how likely it is that the observed difference between the two groups arose from pure chance. Suppose for a moment that in fact, the drug does nothing for blood pressure. If each of our groups had 5 people, we can ask: what is the probability that all 5 experimental subjects would exhibit a decrease in blood pressure, and all 5 placebo controls would exhibit no change or an increase? Even if the drug in fact has absolutely no effect on blood pressure, there is a $2^{-10} = 1/1024$ chance that we would get these results. Every year, very many experiments are performed. For every 1,024 experiments like this, one of them, on average, will get a "false positive" result, *i.e.* a chance effect that probably wouldn't hold up if the experiment were tried again.

These are just a few of the many issues that need to be taken into consideration throughout experimental design, data analysis, and interpretation. The proper language for data analysis is statistics, and the best statistics can do is provide the *likelihood*, given the data, of false positives and false negatives. Nothing is iron-clad…it's merely more likely or less likely. If you are looking for certainty, science won't give it to you. Religion will, but it's a check that you had better not try to cash.

As with many things, there is a caveat. The lack of certainty concerns claims about universal truths. The blanket statement "This drug lowers blood pressure" is really a universal statement. "This drug lowers blood pressure…for everyone." Without actually testing every single person in the world, we can never know that for sure. All we can do is try to find a representative sample of people and make an informed inference. No certainty there. In contrast, we *can* approach certainty more closely if the claim is *existential*, rather than universal. For instance, if someone says, "There are no white crows," we can prove them wrong by finding a single white crow. Done. What we cannot do is the converse: we cannot prove that all crows are black. The difference here is that the truth value of "no white crows" can be determined with a single experiment, while "all crows are black" requires an infinite number of experiments. For that we would have to look at every crow in the universe, and there's no way to know if we've really seen them all. Maybe a white one will be born tomorrow.

So there is a context in which "scientifically proven" does make sense, but this is not the context that is normally referenced when people use the phrase. But this is only one of the reasons that scientific findings are generally considered tentative rather than certain. Another reason is an abiding and conservative desire, on the part of scientists, to avoid overconfidence.

When Newton formulated his laws of mechanics, the amount of sense it made of the physical universe was completely unprecedented. The laws were so good and so compact that it was considered by some to be the end of physics. What else could be done except resolve a few fine points, do a little mopping up? As it turned out, there was quite a bit more than mopping up to do, though Newton could scarcely have known it at the time. As other scientists continued their work in the decades and centuries after the publication of *Principia*, new results appeared, results inexplicable in Newtonian theory. Ultimately, these findings led to Einstein's theory of relativity. Although built on radically new foundations, this theory included Newtonian physics as a special case. In particular, when objects move very slowly relative to the speed of light, the equations of special relativity look almost exactly like classical (Newtonian) mechanics. Only when speeds become close to that of light do the quantitative refinements introduced by Einstein come into play.

Had Newton's theory been considered "scientifically proven," there would have been nowhere to go. Only by *doubting* the theory was it possible for Einstein to derive a new result that helped advance our understanding. By virtue of the same tentative acceptance, special relativity itself was replaced and will probably be replaced again.

If you don't have a deep understanding of science (or even if you do), one of the most difficult things to get right is the balance between the opposing forces of skepticism ("we know nothing") and confidence ("we know a lot"). For example, a television program I once saw profiled prominent scientists' views on what an alien invasion would be like, if it ever occurred. One point of consensual agreement was that Hollywood had largely gotten it wrong: the odds of aliens having two legs, two arms, and a head are overwhelmingly small, and the chances that the conflict would resemble conventional warfare, only with laser blasters, was judged also to be very small. More likely, said the scientists, the conflict would be much, much worse than is portrayed in films, about as lopsided as that between a human and an anthill. A companion with whom I watched the show was frustrated with the smug confidence of these scientists, saying that we knew nothing at all about aliens; they could be absolutely anything. They could be "sound waves" or "energy beings."

How could the scientists presume to know anything whatsoever about how such a conflict would proceed?

This frustration leans a little bit too far toward the under-constrained "we know nothing" view, while Hollywood leans too far toward the over-constrained "we know a lot." If aliens arrive on earth, will they have mass? Very, very likely. Is it *possible* that they will be massless? Yes, it's possible, but we making guesses here, and it's a safe bet that they'll have mass, if for no other reason than they will presumably want to interact with us, for better or worse. If they have mass, that means that they travel (mass moving through space). If they travel, presumably they travel together. That means ships. Could they "beam" themselves? Sure, but they'd have to beam themselves to a receiver, and the receiver would have to travel...

By reasoning in this way, we can make some guarded but plausible forecasts about the likely nature of such an event, even though it is completely outside our experience. Could we be wrong? Could the aliens be "energy beings" or "neutrino-based life forms traveling through hyperdimensional wormholes," or some other equally incomprehensible thing? Pending a clearer idea of what those possibilities mean, yes, they could be, but again, we're only placing bets. This is science fiction, but it's supposed to be *hard* science fiction, *i.e.* speculation that is neither indecipherable nor known to contradict the best available theories. The same style of thinking, dancing between constraint and imagination, should be applied to all unusual claims, whether they be religious, supernatural, or perfectly ordinary.

I once had a friend tell me about a species of frog that, he had read, could do a standing long jump of forty feet, without gliding or other trickery. No way! I said. Your homework assignment: was I justified?

The burden of proof

> I contend that we are both atheists. I just believe in
> one fewer god than you do. When you understand
> why you dismiss all the other possible gods, you
> will understand why I dismiss yours.
>
> – Stephen Roberts

Despite what I've said so far, the actual lifestyle separation of skeptics and the faithful is not as great as it appears. In fact, we are all natural skeptics about most things; it's just that, somehow, religion convinces some people to replace their native skepticism with credulity when it comes to certain claims.

If I say, "I can fly like superman," you would be justified in not believing me. In fact, you flat out wouldn't. But if I say "I had oatmeal for breakfast," you probably would. Why? Because you know that the correct standards to apply when assessing the likelihood of a claim are (1) The person making the claim has to supply the proof, and (2) the more extraordinary the claim, the more extraordinary the evidence must be.

The first of these standards, that the burden of proof is on the claimant, is something we all apply automatically – except when the person making the claim happens to be from our particular religious sect.

> Dick: "I can fly like superman."
> Jane: "Prove it."
> Dick: "I don't have to do tricks to prove anything to
>   you. Superman is great."

Jane: "Well, you can't expect me to believe such a ridiculous thing unless you give me some kind of proof."

Dick: "Honestly I don't care whether you believe or not. But if you did, wouldn't the world be a lot more interesting?"

Jane: "That's not the point. There are a lot of ways the world would be more interesting. If I had a million dollars in the bank, that would be interesting. But it's just not true."

Dick: "It's true if you believe it strongly enough. Belief is the strongest power there is."

Jane: "I can believe whatever I want, but there's an external reality out there. Namely, the balance in the bank."

Dick: "I feel sorry for you. Needing evidence just shows how cold and empty you are inside. If you have the courage to believe, your heart will be filled with love and grace, and you will realize you don't need *real* money."

Jane: "And I suppose you don't need a real plane to get to London."

Under the second standard of extraordinary claims requiring extraordinary evidence, demonstration of actual superhero flying ability would need to be accompanied by proof that there were no hidden wires, no camera tricks, no hidden propulsion device, *etc.* Even if I satisfied all those criteria and still delivered the goods, you would, if you were the curious sort, probably still want to know more about how it is that I manage to do it. Gravity still affects me (the vestibular organs in my inner ear work properly, even in flight), so I must be exerting an upward force to counteract it. Where does that force come from? Is it generated by my muscles? Given that muscles work by moving actin and myosin past each other, how does that force exit my skeletal frame? And what does the force push against? The air? If so, does the grass under my feet flatten upon takeoff? Extensive investigation is required not just because it is an unusual claim, but also because the ability seems to contradict many well-established principles. Whether you have formulated those principles in terms of physics is immaterial – if I can really fly like superman, someone is going to have to reformulate our understanding of why cups fall off tables, why DC-7's can go from Chicago to Dallas, and probably why the Earth orbits the sun. All these phenomena are connected. If we're wrong that superman-like flight is impossible, then we're wrong about everything else too. None of this rules out the legitimacy of my claim, but you darn well ought to be skeptical about it.

If your evaluation of a superman claim moves along these lines, why would you forfeit those same commonsensical demands for evidence when it comes to equally surprising avowals of even weirder things like deities or ghosts? Only because you want to believe them so badly that you're willing to suspend disbelief. I sympathize! There's no question that mysteries can be fun, and the feeling of awestruck wonderment encourages humility and curiosity. But what vexes many scientists is that such thrills are plentiful in the study of the natural world, and they carry the additional satisfaction of real discovery if one is a sufficiently patient and clever investigator. The mysteries of religion and X-Files pseudoscience are, in contrast, ends in themselves. The fun ends as soon as the claim is made. You saw Bigfoot? Wow! How about them Sox? Meanwhile, all reasonable scientists would agree that our natural inquiries are far from complete. Practically speaking, there will always be new things to discover and new ways of conceptualizing natural law, so if it's mystery you're after, *science*, not religion or UFOlogy, is the place to be.

On being a scientist

Skepticism and faith are, I have argued, irreconcilable. These two stances are, respectively, the methodological cores of science and religion, and I think that the two are not just irreconcilable, but antonymic. This will no doubt draw fire from those who point out that there have been and still are many prominent and productive scientists who believe in God. It is true, as far as it goes: religiosity does not prevent a person from having a successful scientific career. But the idea that the coexistence of the two modes of thinking in one person somehow implies conceptual harmony is so specious that we should really make a distinction between (1) having a successful scientific career and (2) *being* a scientist.

If someone goes to church once a week, and participates in the customs of the denomination, but believes not a word of it, does this make them a member of the faith? No way. Conversely, if someone is deeply committed to a faith but does not attend church or participate in its rituals, are they a legitimate believer? Of course. Being a believer is about how and what you think, not solely about what you do. Actions are (in most sects) neither necessary nor sufficient for the identity; thoughts and beliefs are. It's the same thing with science. It is perfectly possible to participate in the actions of science – submitting grants, formulating hypotheses, designing and conducting experiments, analyzing data, writing papers – without being ideologically committed to a single shred of what has gone on. I like to think of people who do this as domain-specific zombies, though a more charitable evaluation of someone who mechanically executes the rituals of an ideology with which they disagree might simply be hypocrisy. More important than the choice of slur is the educational process that makes such people possible.

Undergraduate training for a scientific career focuses primarily on the facts and concepts of a particular field. Graduate training deepens the overall knowledge, begins to build an area of specialization, and imparts the skills necessary to actually become a "primary investigator," *i.e.* someone who runs their own lab. Running a lab, in turn, as a professor or private sector scientist, can encompass administration, technical training and supervision of students, managerial and conceptual oversight, and, of course, classroom teaching. In most cases, these responsibilities leave PIs little time for the hands-on work of science, which is instead done by graduate students and postdocs. And in anyone's hands, experimentation and even conceptual oversight *can* be (though certainly don't have to be) fairly mechanical processes. With a suitable history of research, an unturned stone is easy to find, a conceptual gap located, an experiment designed, and a student supervised. Voilá: scientific autopilot.

I certainly don't mean to imply that this behavior is the norm. There's no telling how common it is. The point here is not to cast aspersions on sincere and committed investigators, but to show how it is possible to work productively in a scientific profession and yet be disconnected from its deepest principles. Whether those principles were *ever* understood, at all, is yet another problem. Conspicuously missing from every stage of scientific education is a program of philosophical integration. Students are rarely if ever asked to place their learning into a larger, coherent worldview. Why this is the case is largely a matter for conjecture, but there are probably several reasons.

First of all, how do you test whether someone has a coherent worldview? It's a much fuzzier thing than whether they know how to run a gel or do a serial dilution. Worldview is really a much softer, much more personal issue than those with which science education typically deals. Insisting that someone *believe* the naturalistic worldview seems perversely fascist, even though we are very comfortable putting equal insistence on knowing the multiplication tables. That may be because we think of worldview not as a fact, but as a matter of individual opinion. The naturalistic worldview, though, is not like other worldviews. It is just the assembled methods and findings of science, and we *do* insist that students learn those. The real problem here is that scientific disciplines have become so overspecialized that no one – really, no one – gets a good grounding in science *overall*. This enables the production of well-qualified cognitive psychologists who know almost nothing of physics. Or of well-qualified physicists who know almost nothing of biology. I would hazard a guess that from the Renaissance to the present, the scientists least likely to be theists were those who had the broadest knowledge of science overall, not those who were most accomplished in a particular discipline.

A second problem is that an embrace of the naturalistic worldview is incompatible with belief in a religious worldview, and this puts educators into a very difficult position. Universities don't want to be forced to fail a student because he or she is a theist.[44] The problem, of course, is when theistic beliefs conflict with logical and empirical methodologies that must be understood for the demonstration of competence in a particular field.

Third, never mind worldview…it is probably impossible to know whether someone actually understands a concept. All you can do is ask them questions and watch their behavior. You cannot ever know whether they "get" the idea, or if instead they are a domain-specific zombie, treating it like a game, limited in scope, by whose rules they will abide for the purposes of amusement and accomplishment. If they say the right things and do the right things, how can you tell – and does it matter – if they really understand? They have the tools they need to apply for and receive grants, design and conduct experiments, and contribute to the body of the scientific establishment.

Fourth, and finally, the machine of education has produced professional students who are excel at scholastic achievement, but this does not necessarily mean they are good at anything else. One of my most vexing recollections in this regard comes from graduate school, during a study session with fellow students. We had become stuck on a particularly thorny mathematical procedure, and were burning valuable time with little progress. From prior examples, we knew how to get the right answer should the question appear on the exam, but none of us understood *why* it was the right answer. Finally one of my friends said, "Never mind why. If the question is on the test, this is the answer. Let's move on." This was an eminently practical suggestion, coming, in fact, from a student far better than myself. But isn't there something terrifying here? The haunting legacy is that this is precisely what formal education has produced: students who are rewarded for placing a higher priority on regurgitation than comprehension.

Test-preparation companies will further attest, in support of this view, that test taking is a skill. The makers of the SAT are consistently very careful to point out that their exam does not measure intelligence, or comprehension, or logicalness, or critical thinking ability. What it does do is predict academic performance. Former physics professor Lawrence Lerner writes:

> Students in the introductory level [Physics] course soon find that much of what they must learn is counterintuitive. Very early, they are exposed to Newton's first law of motion, which asserts that a body on which no force is acting maintains the speed and direction of its motion indefinitely. But this conflicts with the experience they had that very morning while driving their cars to campus...Of course, the better students come to understand that the coasting car is not an example of an object on which no force is acting, and they reconcile the two experiences in a consistent manner. Certainly, all students who want to become physicists must do so. But an awful lot of students who solve enough homework problems to pass the course come to believe that the real world and the "physics-class world" operate according to different laws. It is their obligation, of course, to learn enough about the "physics-class world" to pass the course (and maybe to become computer engineers or physicians or X-ray technicians). But they feel no need to reconcile that world with the one in which they

drive their cars and generally live their lives. And many of them never do so.[45]

For these reasons and perhaps others, the process of preparing a student to become a scientist focuses on the most objective and least controversial demonstrable competencies: (1) Do they know the facts? (2) Can they apply the concepts? (3) Are they capable of making contributions to the field? None of these things test whether they have a logically coherent or even empirically defensible worldview. They test what the student can *do*, not what they *are*. So the way the situation is set up makes it perfectly possible for a person to *do* science but *be* a theist.

Does it matter? This is actually a very interesting question in the "Do the ends justify the means?" family. Never mind theists for a moment – suppose an impossibly fast computer / robot became capable both of generating and testing novel scientific theories. Would we be satisfied in accepting those theories as the products of science? I submit that we would accept and use them, and enjoy their beauty if any, but with some measure of regret. This would have nothing to do with science *per se* but would merely be a byproduct of human social psychology. Group solidarity and shared values are emotionally important. The scientific community feels like any other, and its participants are driven by the same emotional forces. Some part of that group solidarity is eroded when the fruits of its labors, *i.e.* theories, are generated either by brute force or by human practitioners who sacrilegiously play along with a methodology they privately believe is just a toy.

Another way that some practicing scientists manage to be theists is by interpreting God in a way that makes Him recede into the corners of scientific ignorance. Because scientific knowledge is always changing, this is God on the run, also called "God of the Gaps." Questions like the cause of the Big Bang are presently unanswered by science, so some theists simply place God there. Of course, if science ever answers that question, God will be evicted from that particular corner...but this will not be the end of belief. God will simply emigrate and become the answer to the next unanswered question. But God is still tied to the evidence; he is just tied in the reverse way as everything else. As Michael Shermer notes in *How We Believe*, using evidence to steer belief in God is not faith at all. It would be truer to the spirit of religion to simply make belief immune to all reason, immune to all evidence, and assert impossible things merely because they feel nice. By definition this will never make any sense – no more than when a racist says they feel better subjugating others – but at least it will be honest. This is not, however, what religion does. Religion says that it is based on solemn truth, not happy fiction.

Is science just another faith?

What about the claim that science is just another faith? Religion is faith in God, science is faith in naturalism and logic. Does this claim hold water?

The rational skepticism of science applies not just to the investigation of natural phenomena, but to itself as well. This takes the form, for example, of insisting on the internal consistency of a set of data or a theory, the demand that experimental procedures protect against bias, and statistical evaluation to determine confidence levels. These forms of self-policing are effective tools to prevent the backward slip of the scientific view as a whole, which is why the breadth and depth of scientific understanding are always improving. In that sense the criticism that science is a faith is a little like accusing someone who insists on fairness of being unfair to the player who wants to cheat. Or like accusing an open-minded person of in fact being closed minded, because they're not open minded enough to accept closed-mindedness. If science's relationship with reason is faith, then it's a strange kind of faith…one whose only tenet is that there should be no tenets. That's just not the same sort of thing as faith in a supreme supernatural being who created the universe and takes a personal interest in the affairs of every human being.

On the other hand, you won't get anywhere philosophically if you don't have axioms, and axioms are a little like articles of faith, in the sense that you can't prove them; all you can do is say, "this sounds reasonable" or "this is productive." One suggestion for the main axioms of the scientific worldview is this:

(S_1) An objective reality exists.

(S_2) Experimentation is the best way to learn about it.

In that vein, someone might argue that there is no more reason to believe the axioms of the scientific worldview than those of a theistic worldview, *e.g.*

(R_1) One or more gods exist.

(R_2) Revealed truth (including the Holy Text) is the best way to learn about Him / Them.

Superficially, these two systems might appear to define similar philosophical foundations, but a deeper look reveals at least two important differences.

First, R_1 is a weightier assumption than S_1. When scientists say that reality exists, all they are really doing is recapitulating Descartes' *cogito, ergo sum* argument. The mere fact that we are even debating whether reality exists or not means that it does (but recall the previous chapter's discussion of the dangerous word "exists"). The axiom does not make proscriptions about the nature of that reality. It could be a schizophrenic one whose laws change from moment to moment, or it could be a "brain in a vat" reality, or it could have different rules for every observer – these exotic possibilities would make the attendant science difficult, but they are not excluded by the axiom. Thus S_1 is a rather minimal claim by virtue of its flexibility in the nature of "reality." In contrast, R_1 makes reference to a God or gods. Whatever is meant by these words is not specified, but it would be fair to say that there is some disagreement from believer to believer about God's characteristics. As Dennett says,

> Even if you believe that the Bible is the last and perfect word on every topic, you must recognize that there are people in the world who do not share your interpretation of the Bible. For instance, many take the Bible to be the Word of God but *don't* read it to rule out evolution, so it is just a plain everyday fact that the Bible does not speak clearly and unmistakably to all. Since that is so, the Bible is not a plausible candidate as common ground to be shared *without further discussion* in a reasonable conversation.[46]

And quite obviously, the would-be God of the Bible is but one among many. Poseidon, Mithras, Odin, and many others would need to brought into the mix, with convincing arguments as to the plausibility and/or productivity of asserting the existence of each. It is the relative unlikelihood of achieving widespread agreement on R_1 *without further discussion* that makes its axiomatic status different from S_1.

Another difference between the sets of axioms is that even on its own, R_2 leads to contradictions, because there are multiple holy texts and they disagree. Knowing how destructive logical contradictions are, something must be done to alleviate this, but because they are based on faith, theistic systems provide no method to discern true claims from false ones and thereby derive a coherent and objective understanding of the world. Indeed, the only really defensible position for faith is one in which logical contradiction is just fine. But in that case our journey ends abruptly, leaving us stuck in a position of knowing nothing – not even the axioms we started with are safe, because another faithful believer can contradict them with equal standing and authority.

The last comparison to make between the axioms concerns *Occam's razor*. Suppose that we think that the chances of the Red Sox winning the World Series this year are one in a hundred, or probability .01. Suppose we also think that the odds of the Patriots winning the Super Bowl are one in twenty, or .05. What are the chances that both things will happen? To answer that, we assume that the two events are independent, and then just multiply the odds of the two events to obtain our answer: the odds are one in two thousand that both teams will win. Converting from odds to probability, we conclude that to calculate the probability of a compound event, we just multiply the probabilities of the component (and independent!) events:

$$p = p_1 \cdot p_2 \cdot \ \ldots \ \cdot p_n$$

Now instead of sports events, consider a set Q of propositions $\{q_1, q_2, \ldots, q_n\}$ whose truth-values we do not know, but whose probabilities we can estimate, however roughly. We know that probabilities are always between zero and one, so the longer the list of propositions, the less likely it becomes that all will simultaneously be true: multiplying a quantity by something smaller than one can only shrink the result. If we were fans of an infinite number of sports teams in an infinite number of sports – truly infinite, not just very big – the odds that every single one of them would win the championship in the same year would be zero.

Occam's razor says that, for these reasons, we should keep our list of assumptions as short as possible and make each one as simple as possible. The reason was given above:

the longer and less obvious the list gets, the lower the total probability. Because the proposition R_1 "God exists" *includes* proposition S_1 "Reality exists," (God exists in some reality or other), R_1 is necessarily more complex and therefore less likely to be true. This doesn't necessarily mean that R_1 is false. It just means that all other things being equal, we should adopt the S axioms instead of the R axioms, and science is, at the very least, less like a faith than religion.

*   *   *

The original question of this section referred to a possibly even deeper "faith" in logic itself. Is logic falsifiable? Not exactly, but the question is a bit of a red herring. As described in the previous chapter, logic is simply a careful distillation of language itself, not an independent axiom. Besides, it is used by everyone, scientists and religionists alike – including those who would use reason to show that the use of reason is unreasonable! What's wrong with the following argument?

- Science depends on logic
- Logic cannot be trusted
- Therefore, science cannot be trusted

Clearly, we're all in the same boat. The only real difference between the faithful and the skeptics, *vis à vis* logic, is that the faithful have decided that they should be allowed to suspend its strictures at their convenience.

As odd as it sounds, using logic to effectively hog-tie itself is in fact possible; it was done by the mathematician and logician Kurt Gödel in 1931. The result was what is almost certainly the most vexing and celebrated proof in the field, an idea known as *Gödel's Incompleteness Theorem*. This theorem says, in short, that there will always be true statements which logic cannot prove. A full discussion is beyond the scope of this book, but a few points should suffice. First, the theorem shows that logic (which I am substituting for "reason") is not exactly a faith, since it apparently can and does inspect its own limitations. Second, the kinds of theorems which are true but unprovable are not of the "God exists and Heaven awaits" type. They are much more plebeian things, exotic by virtue of their status as children of self-reference, which only a mathematician could love. Third, this is logic we're talking about, not experimental science. The constructions that make Gödel's theorem possible are completely symbolic. Before you go off thinking something along the lines of "See! Science admits it can't prove everything! There's space for God after all!" remember that at the end of the day, this theorem effectively amounts to some marks on paper, and marks on paper have never and will never determine what can or cannot happen in the universe. Logic can help us understand what we find out there, but nature has the last word.

A brief digression. The personification of nature, as in "nature has the last word," may strike some as remarkably commensurate with the personification of God. Some may think, in other words, that scientists have deified nature and are therefore religious. This is not so. First of all, personification is merely a convenient way of writing and of conceptualizing the idea that nature, not any construct of the human mind, should be our ultimate point of reference, and my use of that phrasing here does not betray a metaphysical similarity with theism. One may equally say "the water wants to flow downhill but it can't because it is stopped by the dam." Speaking in this way is just a convenient way to construct mental models. Secondly, as I hope to have shown, science is faithless. Religion without faith isn't religion. End of digression.

Physicalism and phenomenology

Another position attributed to science, seen by some as an article of faith, is physicalism. In short, this is the idea that whatever physics finds in the universe, *e.g.* matter, energy, whatever, is all there is. There are no spirits or deities or powers or forces that are not made of *stuff*, and there are no properties of things not explainable in terms of the stuff and the way it is configured. If this is an assumption of science, the argument goes, then of course scientists will never find evidence of God or miracles, because such things are *assumed* to be impossible; science would in that case be effectively "God blind." Does this make physicalism a faith? Said differently, can physicalism be falsified?

By the definition above, it would seem that if anyone could "point to" a thing in the universe that was not a form of mass-energy (limiting our conversation to present-day physics), then physicalism would be dismantled. The problem with this proposal is that the devices that would ostensibly be used to determine the non-physicality of the thing would probably, by definition, not be able to see it, because the design of all present human artifacts is based on the idea that all things are composed of mass-energy. "Looking at" something involves bouncing a photon or an electron or some other energetic particle off of the target, and the only way the bounce is going to happen is if the target has a manifestation in the physical universe. By virtue of that fact, it may initially seem that physicalism cannot be falsified.

Another way to disprove physicalism, however, would be via the demonstration that two objects, identical in every way down to their elementary particles and placed in identical environments, can act differently. Most people assume that if you copied a human being, you would not get an identical human, but that *is* an assumption and cannot be taken as a refutation of physicalism. Moreover the experiment is impossible given the current limits of technology, and might remain forever undone in any case due to ethical considerations. Genetic cloning doesn't even come close to perfect duplication because of the myriad complexities of development and change throughout life, not to mention complexities of the environment. But even if perfect duplication of a person or a cat were possible, so complicated a demonstration is not necessary. Physicalism says that *any* two objects, no matter how simple, should behave the same if they're built the same. To the extent that we are able, we have done such experiments countless times, and to the limits of our ability to measure, the results always support physicalism. Two hydrogen atoms behave the same way given the same experimental setting. Multiple collections of trillions of such atoms also behave the same. When analyzed statistically, even vastly more complex arrangements of matter that are not strictly identical, such as mice and automobiles, behave the same when conditions are controlled. If such widespread consistency did not exist, we would be helpless to influence it, and all our technology would be lame.

So it is true that scientists are physicalists, but in the same way that they are evolutionists. Physicalism is a conclusion based on observation. It is a *finding* of science, not an axiom. As such, it is both tentative and falsifiable, and could be proven false at any time. Candidly, however, we must admit that such experimental proof, if it came to light, would be viewed with considerable surprise. Because of its profound ramifications, skeptics would be justified in requiring especially strong demonstrations that the two objects and environmental conditions really were precisely identical.

Some philosophers of mind claim that they have no problem finding something that is not physical. Such people, *dualists*, say that their mental experiences are not physical, even if they *originate* in something physical, *i.e.* the brain. In other words, a dualist would say that her enjoyment of strawberries is not *itself* a form of mass-energy or space-time or anything other material thing, because it cannot be pointed to by scientific instruments. Even if a neuroscientist points to a set of neuron activations proven to be associated reliably with the enjoyment of strawberries, it is still just a collection of neuron activations and does not convey what it actually feels like to enjoy strawberries – and it is that "what it feels like" which constitutes the *sine qua non* of the experience. For dualists, this explanatory gap proves that physicalism is false and science will always remain incomplete.

This is the jungle of the contemporary study of philosophy of mind, from which there may very well be no *philosophical* escape (though there are other ways out). The dualist's argument is convincing to many, possibly because our internal experience – our "phenomenology" – is so immediately compelling. To see what is wrong with the argument, consider the following thought experiment.

Modern video games are so sophisticated that it is convenient to talk about one object being larger or heavier than another, because the software of the game simulates its own private physics. Despite our ability to talk intelligibly about one object being larger or heavier, however, these statements make sense only within a limited universe of discourse, *i.e.* that of the fictional game world. In the "real" world, the objects of the game do not have a size or a mass or even exist as objects. When we say the "real" world, we mean the world outside ourselves, the world in which our bodies move, the world that contains the earth and the sun. In that world, the objects of the game exist not as objects, but only as patterns of bits inside the CPU of the game console. But in that case – and this is the crucial point – our interpretation of statements about the comparative masses of the game objects is quite different than the interpretation of similar statements made about real world objects. In particular, we would *not* say that in the real world, one in-game object was larger than another. Instead, we would accuse a person who said such a thing either of talking nonsense or of exhibiting some odd sort of confusion about what constitutes reality.

It is much the same with statements about the phenomenological universe. We can discuss in rough-hewn and practical terms such things as whether an apple is red, but in the end, that discussion is about the phenomenological universe, not the "real" external one. We often ignore the distinction because it is convenient to do so, and because we evolved in settings where there was no reason to make it. But as Wittgenstein noticed, the rigors of philosophical discussion snap us out of our linguistic lassitude and make us attend more scrupulously to the real range of our statements. If I say that I know what it is like to enjoy strawberries, and that this feeling differs from what it is like to see the color red, this is a perfectly sensible thing to say. Both statements are being made with respect to the same universe of discourse, *i.e.* the phenomenological one, and can be reasonably compared. But if I say that I know what it is like to enjoy strawberries and that a description of neurobiological dynamics is insufficient to explain it, then I am attempting a trans-universal comparison. It should be no surprise that one cannot account for all the specifics of the other; it is the same as if I complained that I knew – within the context of a game – what it felt like for an object to be too heavy to lift, and that this was not the same feeling as I had when I contemplated a description of the software that instantiated the game's relevant physics. It is not a weakness on the part of the game programmer that his description of object weight lies in discussions of subroutines, nor is it a mystery why my *feelings* would not be addressed by that description. There is nothing unusual happening here. No explanatory parity could ever be achieved via such comparisons, because the phenomena being described live in different universes. Neurobiology lives in the world of the earth and the sun, and the experience of enjoying strawberries lives in the world of idea.

Dualists may retort that the phenomenological universe is *real*. What is argued for above, they would say, is merely the incompatibility of explanations across universes of discourse – there is no dissolution of either universe. But in asserting the reality of a phenomenological universe, they are claiming the existence of something and asserting that it has the same kind of existence as the sun and the moon. Only if you have an outrageously liberal understanding of existence and reality – one which would require you also to admit the reality of Santa Claus and the tooth fairy – would that claim be convincing.

Science wishes to describe the world. What is the world? It is the objective reality that we all reasonably agree exists, even if our experience of it is subjectively filtered. Although it makes sense in our conversations to grant some type of "existence" to anything that can be thought, we are under no obligation to casually accept that that type of existence is the same as the type of existence possessed by everything else in the world, *e.g.* rocks and trees and photons. If I dreamt that the tooth fairy stole my retainer, the fairy's crime is not an event in the natural universe...it is an event in the phenomenological universe, *and that entire universe exists only at the pleasure of the underlying biology.* Enjoyment of strawberries is contingent on a collection of neural events. What is denied by physicalist neuroscience is the demand, made by phenomenological prisoners, that the events inside their minds *exist* with the same sort of reality as the physical events that gave rise to them. Such demands have no jurisdiction in a discussion of "free processes," *i.e.* those outside the phenomenological cage.[47] In other words, only if one fully understands what is meant by the word "exists" can one intelligently say that the enjoyment "is" anything at all.[48] If enjoyment fails to "be" anything then it also fails to dictate the range of science's application. Thus the philosophy of mind is really the philosophy of language. But as I showed in Chapter 4, the philosophy of language is a usurper king – the true king is nature, embodied here as the brain that gives rise to phenomenology and its associated philosophy in the first place. Science does not explain what exists. Science explains nature, and existence is not a part of nature. Existence is a phenomenological construct.

"Do you believe that absolutely everything can be expressed scientifically?"

– Hedwig Born to Albert Einstein

"Yes, it would be possible, but it would make no
sense. It would be description without meaning, as if
you described a Beethoven symphony as a variation
of wave pressure."

– Albert Einstein

Einstein here voices a position with which the dualist would
likely agree, *i.e.* the centrality of "sense" and "meaning." As
we have seen, the whole point of neuroscience is that its
ultimate extension will include a physicalist account of those
terms and the phenomenology they entail. Sense and meaning
are symbolic and linguistic terms. Where does language
happen? In the brain. Strawberry-liker and Einstein alike have
simply hypnotized themselves with language. Once the
neurobiological origins of language are understood, such
complaints evaporate like dreams upon waking.

*   *   *

This has been a challenging and pointed discussion of
methodological conflict, but an important one. The issues
addressed in this chapter include many of the most
foundational differences between naturalistic and religious
ways of thinking. Knowing specifically why the two are not
compatible is central to avoiding a noncommittal pick-and-
choose attitude. Science and religion are not working from the
same script. Their ways of guiding belief are completely
different. More to the point, the religious way of guiding belief
is so deeply flawed as to be worthless…actually, worse than
worthless, because there are infinitely many more false
proposals than true ones, and faith grants them all equal
merit. This is emphatically not to say that religion as a whole
is to be dismissed – only that its methods of discerning true
from false are unreliable and cannot be used to understand its
overlap with the claims about nature made by science. If we
are to really understand religious experience, or anything else
for that matter, we cannot do so with religious methods.

In addition to those weaknesses of faith as a method, we have also seen how the semi-religious claim of dualism and the constraints of Gödel incompleteness ultimately fail to dethrone the scientific method as the best way to learn about the world. If this was your first exposure to Gödel or the philosophy of mind, the take-home message is that anchoring supernaturalism in the mysteries of the phenomenology or the Incompleteness Theorem ultimately doesn't work. Nature may force us to confront some bizarre phenomena in the end, but if so, those phenomena probably won't have anything uniquely to do with human beings, whether it be where we came from, how we think, or what we do. If there is something physically special about us – and science is open to that possibility – demonstration of that is going to have to run the gauntlet of requirements specified in this chapter.

Humanism

Possibly in partial response to religious-based entitlements, some secularists have developed a philosophy that favors humanity but does so without recourse to anything supernatural. While the placement of our species into a defensible ecological context may be more informed, however, the anthropocentric stance of secular humanism is still apparent. In some ways it is even more insidious, because secularists use their pro-humanity positions as markers for morality. To combat the idea that one cannot be moral without God, secularists take the defensive position: "Look how moral we are: we love people just as much as religionists!" This conflates morality with anthropocentrism.

Our self-interest is so pervasive that it is embedded in our language and exposed in the way we talk and think. For something to count as progress, we generally require that it produce a benefit for humanity. Perhaps it's natural that the "humanities" would be anthropocentric – that artists, historians, and writers would be primarily concerned with the works and deeds of other *Homo sapiens* – but is it obvious that the sciences would also be self-obsessed? In principle, scientists study things so we humans can have a better understanding of the world around us. In practice, though, there is more grant money available for applied than pure science, and if you press them, many scientists feel pressured to give you *some* useful application for their studies – quite often it's not about the knowledge *per se*. Chemists work to find new substances with desirable properties (*i.e.* desirable to business); biochemists search for medical and industrial applications. Environmental policies, which ostensibly are about the environment, are often designed to save wilderness so that we can enjoy it, to set aside parklands so we can hunt, hike, and camp there, to preserve tracts of forest, permafrost, wetlands, prairies, and ocean just in case we someday need the resources they contain. All of these are selfish enterprises packaged to look selfless.

This delusion has made its way into our vocabulary. "Humanitarian," "philanthropic," "generous," and "considerate" are almost synonyms of "good," "decent," and "wholesome." And yet, each of these adjectives refers exclusively to our own group and species needs. We vilify racism and sexism and many others isms. Is humanism not the same devaluation of *the other*, this time of the ecosphere?

As ugly as this self-obsession is, perhaps it comes as no surprise. Anteaters, were they sentient and technological, would probably not be any more outward looking than we are. They would mostly be interested in the betterment of anteaterdom. From a Darwinian perspective, one expects that cuttlefish and naked mole rats, when they get their turn at global dominance, will likewise be interested mostly in making themselves more comfortable, their lives longer and more luxurious, their children happier. So it's not that we should blame anyone or assign especial fault to humans. It's just interesting that the attempt to exalt humanity above the rest of the animal kingdom cites our sense of ethics as a key factor, while ethics, in most cases, turns out to be unvarnished narcissism. Since when is looking out for number one while trashing the rest of the world a sign of nobility?

*   *   *

Suppose you discover that you have a serious but operable tumor near your brainstem. In a meeting with the surgeon who is to perform the operation, you notice a plaque on the wall, prominently displaying the following guiding principles:

1. Surgeons are better than patients. Our welfare comes first.
2. Easy come, easy go. If the patient dies, at least their suffering is over.
3. If you nick an artery, don't sweat it. You're not responsible, God is.

Any religion heralding humans as the unique inheritors of God's favor endorses, implicitly or explicitly, at least some of these ideas, suitably translated. Humans are closer to God than other animals. Life on Earth is just a holding pattern awaiting the Kingdom of Heaven. Whatever happens, it's for the best, God has a plan. These are not purely Old Testament ideas; most are features of many faiths. And yet, it would seem that, at a minimum, the *exact opposites* of these ideas are more likely to yield a responsible ecological stance:

1. Humans have unique traits, but so do bees and bats. We're not "better than" other animals.
2. Life on Earth is all we, or any other animal, get. Heaven is ours to build, but it has to be right here, right now.
3. Be careful! Bad things happen all the time, and extinction is for keeps.

What's odd is that when reduced to its essence, this second set of principles seems like it could almost fit into the tenets of a post-religious but quasi-Abrahamic ideology. Humility. Personal responsibility. Something resembling penitence. But when ideas like these are endorsed by religion, it is with a domain of application hamstrung by an accompanying belief in the divine origins of our species; our souls, if not our bodies. Do you really expect that the plight of the pygmy burrowing owl will evoke feelings of personal responsibility from someone who thinks that an omnipotent all-knowing creator of the universe invested them, but not the owls, with an immaterial soul? Someone who believes that the Lord Our Savior controls all things, causes all things, and has a plan for the universe? That soon, Jesus will return and take all believers to eternal bliss in a heavenly afterlife? If one small population of dust-covered owls is silently snuffed out of existence, somewhere in a desert that lacks even an observation platform for a few amateur ornithologists to gawk at them (what else are owls good for?), is that really going to hold back the glorious tide of redemption and peace that God has planned for his beloved children?

Rhetoric aside, the actual relationship between religiosity and environmental attitudes has been studied. In a 2004 study by the Pew Charitable Trusts, the subtitle stated what the researchers consider a central conclusion: that environmental support is strong across a wide variety of religious groups. That is what the title says, but not the data. For example: the survey probed what percentage of people, within each particular religion, agreed that we need stronger environmental regulation. This number was never 100% – some subset always disagreed or had no opinion, so it is appropriate to ask what the margin was between those agreeing and disagreeing. By that measure, the group most in favor of stronger environmental regulation was atheists and agnostics. 66% were in favor, 11% opposed, giving a margin in favor of 55%. In contrast with this, consider the largest single religious denomination in the United States: Evangelical

Protestants. The Pew study indicated that fully 26% of American citizens belong to this group. Among them, 52% were in favor of stronger environmental regulations, and 31% were opposed. With a marginal support, then, of 25%, Evangelicals were significantly less than half as likely as atheists to support additional environmentally protective legislation.

Another finding of the study supporting the anti-environmental attitudes implicit in religious worldviews was that, across all faiths, the more traditionalist sects were less supportive of environmental regulation than the modernist sects. In Catholics, for example, marginal support was 26% weaker in traditionally-minded believers than in modernists.

Perhaps worst and most bizarre of all, especially in terms of domestic political reality, was the portion of the survey that examined voting priorities. In this question, respondents were asked to prioritize a list of issues – terrorism, the economy, the environment, and so on – that they planned to use to determine their voting preferences. Among white Evangelicals, the environment was dead last. When it came to voting, this group – the dominant religious affiliation in the most industrialized and economically powerful nation on earth – literally cared more about banning gay marriage than they did about *saving the planet*. It would seem that speculation on the putative environmental views endorsed – or rather, *not* endorsed – by religion may not be, after all, entirely off-base.

A more liberal religious view might be that nature is God's creation, and on that basis we should respect it and cherish it. Though this is supported by some of the Pew results and by anecdotal information available from official church statements, it is a weak answer to the hard fact of beliefs as they manifest in practice, and, at best, a disingenuous attempt at environmental stewardship. To say that our divine Father wants us to care for the lesser figments of His creation is to promote the same feelings that you might hold toward the crayon artwork of your second-grader's classmates. Whatever that feeling is, it is not respect. It is more like courtesy toward the cute, more like smug superiority than honest esteem. We praise it with words and find it adorable, perhaps, but not up to our level. *Honest* respect, *real* respect and real reverence for nature is, like other forms of love, humbling, even overpowering. Look at a beetle and think of its sensitivity to natural forces you cannot even detect. Think of its connection to deep evolution, to its ancestors 100 million years ago, and to you as well; the inconceivable speed and precision of its cellular processes; its ability to survive in an immense and incomprehensible world. It sees you and stops, waving its feathery antennae to collect information and analyze you. Can you smell a difference between your left hand and right hand? It can. Can you see yourself in infrared? It can. Somehow, its nervous system knows how to process this information and generate appropriate action plans. Run away? Eat you? Try to flip you over? The sophistication of this creature is almost unimaginable. Standing before this beetle, deep respect for nature should make you humble to the point of tears-in-your-eyes, nervous, giggling, trembling prostration. Like the God I am encouraging you to dismiss, nature deserves *awe*, not politeness. If you've never seen anything capable of producing this emotional state, you are missing out on the most profound experience of natural glory. Awake and alive, to be sure, but unaware of the Eden we scorn while waiting for Heaven.

The alternative to religiously or dualistically-motivated selfishness, of course, is a set of philosophies, social policies, and most of all, *behaviors* that advance the Copernican, Darwinian, and neuroscientific revolutions. We must accept the truth: the sun does not revolve around the earth. The earth does not revolve around humanity. We are not the favored creatures of God or of the universe. We are an idiosyncratic branch of primate evolution, newcomers to the global ecology. To be truly noble, we should accept the responsibility that accompanies our great intelligence and ability and become stewards of our world. Not so that we can later plunder it, not so that it will be there for future generations to enjoy, but simply as a sign of appreciation. There is value in the preservation of the world for its own sake for the same reason that the happiness of a stranger makes us happy – not because there's something to be gained, but simply because it means the world overall is better off.

Lest this sound like feel-good environmentalism – and the Green Revolution seems to be achieving some traction – realize this: symbolic gestures are not enough. Yes, Yellowstone is glorious. Yes, some states have set aside a few coastal wetlands, and yes, we are getting smarter about recycling and industrial pollution. But it is too easy to say, "Just think how bad things would have been had we not done X." Nature doesn't process counterfactuals. All that nature sees is the actual damage done, not the even greater damage that we might have done but didn't. Yesterday I wanted to bludgeon a lazy checkout clerk but didn't. Does my restraint make the world a better place? No. I only make the world better through the positive presence of good acts, not the absence of bad ones. Hybrid cars have improved fuel efficiency, reducing emissions by 40% compared with a conventional engine. Is this cause for celebration? No – especially not if population growth or economic development produce enough new drivers to offset the per-car gain. Even with a hybrid, 60% of those toxic emissions and greenhouse gasses are still coming out of the tailpipe. And you still have to manufacture the cars themselves. Ever seen a tire junkyard? Remember: without us, there would be *zero* automobile emissions.

Granted, it is impossible to live in modern society and do no harm to the environment. You do what you can, but still, you want your plasma TV, your cell phone, your laundry detergent. I understand! I want those too. Even if we all embraced communal agrarianism, ate flax and trimmed sheep all day, it wouldn't be enough to save the world. What to do? Well, at the end of the day, it is the larger trend that matters. Think about, and do something about, population control. (And no, you can't just lobby on behalf of the cause. Don't just tell other people they should stop having children. *You* should stop having them.) Support educational trends that encourage holistic and naturalistic views of our place in the world. Discourage egocentric behavior in all its forms: exclusive humanitarianism, religion, and consumerism. Encourage technological advances that allow us to live comfortably while having a minimal, sustainable impact on the world. Zero impact is impossible and not worth thinking about, but sustainable impact, where populations, consumption, and emissions are stable is achievable and is what we ought to be working toward. The solution is forward, not backward.

Moral relativism

> When you think of the long and gloomy history of man, you will find more hideous crimes have been committed in the name of obedience than have ever been committed in the name of rebellion.
>
> – C. P. Snow

For many people, moral certainty is a strong selling point for a religious worldview. Civilization is a confusing place, becoming ever more so, with strong clashes between cultures and a huge diversity of personal psychologies. It can be quite frightening to imagine that you are lost in a sea of humanity, with no privileged place for your specific value system.

What a comfort, then, to hear an authoritative decree that your values are absolutely correct, and any conflicting values are absolutely wrong. You're on the winning team! Far less satisfying, of course, would be the weak-kneed consolation that, while your values are not "right" in any *real* sense, at least they make you feel good. That latter claim, *i.e.* moral relativism, is indeed a wet blanket to many people, who choose instead to adopt the worldview that elevates them (or their gods or holy text) to the status of supreme moral authority.

But for moral objectivism to be true, what must we assume about the world? Well, what do we mean by "objective?" Presumably, something like this: "Something is objective if it appears the same to all observers." The ratio of the circumference to the diameter of a circle is objective. Whoever calculates it, that ratio will always be the same. Under moral objectivism, any competent (whatever that means) measurer should be able to examine your value system and determine that it, indeed, was "true," or conversely that every action has a clearly discernible moral value, independent of the observer's personal beliefs.

The π example is objective because π is a *logical* attribute. Its value actually follows tautologically from the definition of circle. But what about *empirical* attributes? Are there objective *empirical* structures? How about the ratio of the masses of Earth and a mosquito? Yes, this does indeed appear to be an objective measurement. If any two observers measured these quantities and calculated their ratio, they would arrive at roughly the same number. Even if they used different standards – pounds, kilograms – the units disappear in the ratio.

So there *is* such a thing as an objective *empirical* structure, and we're not chasing a red herring. What would it mean, then, for the world to have an objective moral structure? As I said above, there would have to be something measurable enabling any observer to determine the moral value of a given event. *Any* observer. If that observer were a Nazi, a priest, an American philosopher, a highly evolved ferret, or a sentient oyster from Andromeda, he/she/it must (1) be able to measure the moral value of the event in question, and (2) agree on the values so obtained.

Let's skip ahead and try to fend off an objection. Those opposed to moral relativism often make an argument like this: "Ok, you say morality is meaningless, nothing has any moral value. So I could kill you and that would fine with you. Right? And I should get away with it, because the law would also be morally impotent, allowing everything. You're just being perverse, because I *know* you would think it was bad if I killed you. And a legal system that did nothing about it would produce anarchic chaos."

Because this objection is so common, I want to answer it preemptively. Let's consider the second part first, *i.e.* the claim that moral relativism destroys civilization by endorsing anarchy.

First of all, there is nothing in the claim of moral relativism to justify a conclusion about what any legal system should or should not do. Even if we stipulated the unlikely claim that relativism would inevitably undermine any possible system of justice, rejecting relativism *as false* would be to commit an Appeal to Consequences fallacy. It would be akin to saying that a particular recipe didn't make chocolate cake, because chocolate cake makes you fat and you don't want to be fat. The recipe either makes or does not make cake; whether it makes you fat has nothing to do with the results of following the recipe. Similarly, moral relativism either is or is not true...if it leads to anarchy and destroys civilization, tough noodles, we'll just have to find some way to *save* civilization. Rather than lying to ourselves to get what we want, let's have the integrity to look the problem square in the eye and figure it out, irrespective of how unnerving the consequences. We'll deal with them as necessary.

Secondly, inasmuch as moral relativism *does* have social consequences, it would be helpful – purely as a pragmatic gesture – to understand how to mitigate undesirable effects. And granted, if murder were endorsed as acceptable behavior, mayhem would probably result. So what do we do? Think of it like this: I enjoy living in groups, and I don't want to be killed, so I will support laws that imprison murderers. This need not be accompanied by a moral judgment – it can be a purely pragmatic act. Murderers behind bars cannot kill me. Moreover, the threat of being locked up (a condition many find undesirable) might dissuade at least some would-be murderers from carrying out their plans. By having such a law ("murder shall be punishable by imprisonment") I achieve both proactive and reactive goals – or at least, make some progress toward them – without making any objectivist moral statement. I don't ever have to say "murder is wrong," and I don't have to remain neutral. I merely act in a way that furthers my preferences. Consequences, but not judgment.

Now, am I *right* in doing so? Am I morally justified in punishing murderers? No. On an individual basis, I am just as oppressive as they are. I am imposing my desires on them, just as they would have done to me. Does this make me a hypocrite? No. It makes me a democrat who believes that the needs of the many outweigh the needs of the few. Most of us want to dissuade and punish murderers, and we can do so, but we should always be clear about what we are doing and why. We must never convince ourselves that *universal right* is on our side unless we have a logical or empirical defense. And until a mathematician proves a theorem or a cosmologist finds invisible moral labels attached to every action, we should build our societies and our moral systems to function as smoothly as possible while always operating within the constraints of a reality that we can confidently say *is* objectively real.

If we accept the dogma of moral absolutism, we also make a grave *practical* error. Again, dogma virtually guarantees conflict. If I were a moral objectivist who said that some moral claims were simply True, end of story, it's a safe bet that others adopting my methods would sometimes have beliefs violently opposed to mine. At that point, we can't talk, we can't reason. We can only fight.

Let's now return to the first part of the moral objectivist complaint that murder is "obviously" wrong and moral relativism is simply perverse in claiming otherwise. What does it mean for murder to be "obviously" wrong? Why is it obvious? Presumably it has something to do with an even deeper moral "truth," *e.g.* that the right to life is sacrosanct. You are "entitled" to your life, and I am "not entitled" to take it by force. Sadly, if we go into the universe looking for entitlements, we will not find them any more easily than we will find moral judgments, so we must delve a little more deeply into the question. What are entitlements? Where do they come from? Seemingly, an entitlement is a right, and a right, in turn, is a statement to the effect that something *ought* to be the case. Everyone *ought* to be allowed to live. Once again, what is "ought?" My dictionary gives four senses:

- Obligation (ought to pay our debts)
- Advisability (ought to take care of yourself)
- Natural expectation (ought to be here by now)
- Logical consequence (the result ought to be infinity)

Let's consider each in turn.

Obligation consists in a social contract between parties, whether driven by generally accepted norms of behavior or a legal device. Both are social pragmatisms designed to make groups function smoothly. So at least as far as obligation is concerned, we could simply base a moral system directly on social pragmatism without the intermediary chain of entitlements to oughts to obligations. The middlemen add nothing.

The *advisability* of an action is an appeal to the possibly negative consequences of *not* doing so. "You will not be happy unless you..." This is an appeal to the private welfare of an individual, which in the case of an exhortation not to commit murder would constitute a way of avoiding incarceration or worse. But there is no metaphysically ethical content behind such a warning. It is again purely a pragmatic issue. If you don't wish for incarceration, abstain from murder. Advisability is a private, personal pragmatism, just as obligation is a social one. Again, this gets us to moral *relativism*, not objectivism.

A meaningful *natural expectation* of "ought" would seem to require something embedded in nature. But if there were something natural about ethical precepts, we wouldn't be here in the first place. We are no more likely to find empirically detectable correlates of "ought" than we are to find that an entitlement has a spatial position or an angular momentum. If we seek such a natural basis for the topic at hand, *i.e.* the presence or absence of life, in hopes of finding objective moral value in that, we will find that we are holding in each hand entities from different universes. In my left hand, I hold the ethical concept of value. In my right, I hold a mathematical / physical description of life (whose objective characterization is, by the way, more than a little slippery). How and why should these things be connected? This is the same problem we started with.

Logical consequence would seem to require that entitlements and/or "ought" statements are a question of formal logic, possessing a rigorous proof from first principles. In order to be logical, it would have to be as tautologous for murder to be bad as for 2+2=4. This seems highly unlikely – at least, without some suspiciously *ad hoc* axioms – but it is the only one of the four senses that would seem to offer any opportunities for a would-be objectivist.

*   *   *

If you let on that you're a moral relativist and a causal determinist who doesn't believe in free will, you may get a retort like this: "So, if I can't control what I do and it doesn't matter anyway, why did you get mad at me when I stole your parking space?" At least two responses are needed.

First, this is an *ad hominem* fallacy. I could be a dissolute, thieving, pedophilic crack addict, but none of these character flaws has anything to do with the truth or falseness of my beliefs. It may speak to the personal consequences of believing them (though the sample size would be very small), but that is a matter separate from whether they are factually correct. If I react harshly to you when you steal my parking space, what you get is a lesson on my emotional dynamics, not on moral relativism writ large or causal determinism. To ascertain whether moral relativism and/or causal determinism *as positions* are correct, we need to look at the content of those doctrines, *i.e.* at their logical structure and empirical support, not at the people who espouse them.

The second response to the retort is more complicated, and raises the question of how we build pragmatic day-to-day behavior guides founded on our core beliefs.

As a causal determinist, it's quite true that I believe that people are not metaphysically responsible for their own actions. At the same time, those actions came from somewhere. Where did they come from? They came from the biomachine that is that person. Why did that machine do those things? Because of the way it is built, the way its structures and functions were changed by experience and life history, and the way circumstances affect it. The *ultimate* cause of the behavior, way, way back, all the way to the beginning, is hidden somewhere in the structure of the universe and the big bang. I may never find the ultimate cause of the behavior; at least, probably not in time to do anything about someone stealing my parking space.

The *proximate* cause of the behavior, however, is to be found somewhere in the inner workings of the person who generated it. Because I am a human being, I share at least some of my inner workings with the person who stole my parking space. To be sure, our emotional hardware is not identical, but I can be reasonably confident that this person feels guilt when they know they have done something "wrong," fear or anger when someone threatens them, gladness and pride when they do a "good" deed. This is not to say that my response to them is reasoned out in this way. In fact, my response is only semi-conscious and almost automatic. But if I were to think it through, I might devise a purely rational plan based on my desire to condition this person into ceasing the behavior that I found offensive. How might I go about that? By generating an aversive stimulus that follows the behavior. This is a textbook case of operant conditioning. If Mr. Parking-space-stealer behaves similarly with a few other people, each time experiencing stress when his victims publicly scold him, perhaps he'll stop doing it. This may not benefit me directly, but that would probably be because we live in a city large enough to minimize the chances of our meeting again. It's quite possible, if not likely, that this kind of reaction logic evolved in archaic environments, when social groups were smaller, and there really were no strangers. In a setting like that, the eruption of another person's fury might be significantly more stressful than it is in today's "I'll never see you again anyway" urban environment.

Again, I probably won't go through this whole chain of reasoning in the moments after my parking space is stolen. Emotion overwhelms me, and I start yelling. The rational process that I *will* go through, however, tells me that my anger response may have practical consequences that I desire...namely, that I may be helping to prevent the deterioration of my social environment. That consequence has positive Darwinian value, so my own social conscience has no problem letting the behavior through. Post-evolutionary philosophy comes to bear when I reflect upon the incident to determine how I feel, deep down, about the personal responsibility of Mr. Parking-space-stealer. Do I, as a practical matter, hold him responsible? Yes, because doing so has desirable results. Do I believe he *is* responsible, when all is said and done? No. He stole my parking space because his frontal cortex and amygdala and whatever else had been filled with experiences that told him it was the best overall move. He didn't "decide" to do this on his own...he is a machine doing only what it has been built to do. He is a calculator, he is a mechanism. It's just that I want to change his wiring to make him compute something else.

For one accustomed to notions of naïve and objective morality, this mechanistic take on human behavior may seem cold and calculating, but its ultimate results are anything but. In the first place, it teaches that the motive force of ethical judgment is based on mutual understanding. If I know that Mr. Parking-Space is taking the spot because he's rushing to the hospital to see his injured wife, I might not get angry. And for all I know, he may have just such a reason. If he doesn't, and he knows that I do, then perhaps he will drive away and park somewhere else. Or maybe he won't. Maybe he is simply a rude person. But why is he rude? Maybe he's rude because his colleagues at work belittle and ignore him, and that rejection has turned into outward impatience and hostility. If that's the case, then is it the belittling colleagues who are ultimately at fault? Well, why are they belittling? The chain of reasoning goes back, and back, so far back that we'll never find its source – at least, not until the cosmologists are done with their Big Bang studies. The point is that the moral view of others that we arrive at by considering the source of their behavior is, while mechanistic, still motivated by a deep shared interest in social harmony. That respect for others and our shared interests transforms knee-jerk outrage into a sympathetic inquisitiveness that is – dare I say it? – Christian.

EPILOGUE

Ⅰf there is one question motivating this book, it is actually not about religion or art or philosophy, or even science. It is this: Why do we look for beauty? I hope to have demonstrated at least the plausibility of the idea that in addition to the aesthetic experience, the religious and scientific experiences also are bound together in the neurobiological answer to that question. Further, I hope to have encouraged you that even if the theory presented here is wrong, we have every reason to expect a naturalistic explanation that unifies these phenomenologies.

It may have dawned on you that certain aspects of this book are self-referential, or rather, auto-explanatory. The investigation of why we search for beauty is itself a search for beauty, so the quest has a conceptual closure like a pair of mirrors turned to face each other, creating a new volume. It is similar in some ways to the recursive structure of self-reflective consciousness: by including ourselves in our world model, something qualitatively new seems to arise, as if from nowhere.

As wondrous as these spaces are, they are not magical. They are grounded in the neurobiological, psychological, and evolutionary framework of the brains that make the entire project possible. The reluctance to accept that, the inability or unwillingness to suppress our fear, let go of our talismans and look nature calmly in the eye is probably part of the reason that supernaturalism in all its forms, theism and dualism particularly, persists today. No doubt, when the pharaoh Akhenaten tried to convince his polytheistic citizenry that there was but one true God, he faced an equal resistance. In that sense, it is more than a little ironic that the obstacles faced by monotheism and naturalism would turn out to be so similar.

Steve Martin once quipped that if you want to play a really mean trick on a kid, whenever you're around him, talk wrong. Then, on his first day of school, he'll raise his hand and ask, "May I mumble dog face to the banana patch?" Although I have pushed the envelope in the hope for unification between science, religion, philosophy, and everyday life, I hope I have not "talked wrong" about anything. I have tried to build the foundations of this theory on reason and experimental data, and whenever possible, to make predictions that can be tested and falsified, as is required for any scientific proposal. Where this has not been possible, I hope at least to have constructed something beautiful.

APPENDIX A

Information Theory

Shannon imagined a device (*e.g.* a transmitter at one end of a communication line) that produces one symbol per second from the set {a, b, c}. We don't know how the device works internally, so we are uncertain about which symbol will appear from moment to moment. Each time a symbol does appear, our uncertainty in that moment is reduced, and Shannon said that as a result, we have obtained *information*. How can this be quantified? There are three symbols in the set of possible transmissions, so perhaps we could simply say that this system can transmit one symbol per second and has a information capacity of three symbols. Does this scheme work in general? Let's try to extend it.

Suppose we add a second transmitter – whose mechanics are also unknown – which has a different symbol alphabet: {1, 2}. This second machine also produces one symbol per second. If we combine the two machines into one big machine, we see that there are now *six* possible compound symbols, any one of which might be produced in any given second (a1,a2,b1,b2,c1,c2). If we were simply to add the information capacity of each machine, there would be a capacity for *five* symbols {a, b, c} + {1, 2}, not the actual six. Thus, if we want our measure of information to be the sum of the information produced by each machine (an intuitively appealing as well as mathematically convenient result), we will need a different definition.

Shannon realized that unique properties of the logarithm might fill the bill. Because $\log xy = \log x + \log y$, we can define the uncertainty $U = \log_2(M)$, where M = number of symbols. With the logarithm base 2, the units are then "bits" (short for 'binary digits'). To test the consequences of this definition, notice that with a 1-symbol device, $\log 1 = 0$ bits: because only one symbol can appear, there is no uncertainty. With a 2-symbol device, $\log 2 = 1$ bit. With a 4-symbol device, $\log 4 = 2$ bits. Turning the equation around, we see that if we have two "slots" and can put a 1 or 0 in each, there are four possible strings: 00, 01, 10, 11. Those four strings can encode the original symbol alphabet {a,b,c,d}. The same procedure can be used to encode any size alphabet merely by adding more bits; and adding symbols means adding uncertainty. So far, this seems to make good sense.

But what if the symbols are not equally likely? We said that we didn't know how the symbol-generating machine worked internally, and it is far too strong an assumption to say that all symbols are equally likely. So what if they're not? To handle this possibility, we start again with Shannon's definition but rearrange the system using a trick of logarithms and the definition of probability:

$$
\begin{aligned}
U &= \log(M) \\
&= -\log(1/M) \\
&= -\log(p)
\end{aligned}
$$

where p is the probability of a particular symbol. Probabilities are always between zero and one, so the lower and upper bounds for U are $-\log(1)=0$, and $-\log(0)=\infty$. Thus U, which we have been interpreting as uncertainty, is small when the probability of a symbol is high, and large when the probability of a symbol is small. If we then reinterpret U as a measure of "surprisal," *i.e.* how surprised we are when we get the symbol, this makes perfect sense. The less likely a symbol, the more surprising it is.

Now suppose we have a set of symbols, all mutually exclusive so their probabilities sum to one. The "surprisal" of the ith symbol is then

$$u_i = -\log(p_i)$$

Shannon then supposed that he had a device constantly spitting out such symbols from its alphabet. How can we characterize the amount of information generated by this machine? First, we define uncertainty as the *average surprisal per symbol*. Each symbol in the stream of output from the machine produces some surprisal u_i. Our procedure is to find the average surprisal over all the symbols, and use that one number to summarize the information produced by the device. We begin with a finite message of length N generated by the device. If we select an arbitrary symbol, say the ith one, its individual surprisal is u_i. Suppose that that particular symbol appears N_i times over the course of the entire message. To obtain the average surprisal, we simply add the contributions from all the symbols, then divide by their total number:

$$H = \frac{\sum_{i=1}^{M} N_i \cdot u_i}{\sum_{i=1}^{M} N_i}$$

But the denominator is just N, the length of the string. We can rearrange to obtain

$$H = \sum_{i=1}^{M} \frac{N_i}{N} \cdot u_i$$

In an infinite sequence, the frequency (N_i/N) becomes the probability p_i, so our average surprisal for the whole string becomes

$$H = \sum_{i=1}^{M} p_i \cdot u_i$$

And finally, substituting from above for u_i, we get

$$H = \sum_{i=1}^{M} p_i \cdot -\log(p_i)$$
$$= -\sum_{i=1}^{M} p_i \cdot \log(p_i)$$

This is the average surprisal produced by the symbol generator. This is the central formula derived by Shannon, who called the quantity H *entropy*. It's the same thing as uncertainty.

Now, what is the relationship between uncertainty and information? We said above that information is the *removal* of uncertainty. So suppose – to take a conventional application – that a computer has a collection of on-off (*i.e.* 0,1) memory registers, and we are ignorant of their initial state. In other words we have a device with two symbols in its alphabet and a string of length n. This gives some entropy H_{before} for the existing state of the memory. Then we do something to the computer to set some of those memory locations to values that we do know. Now there is a new entropy, H_{after}, corresponding to the new situation where we have no uncertainty about some of the registers (because we just set them). In this process, we have stored (H_{before} - H_{after}) bits of information.

We're used to applying information theory to digital computers, but how about an example in a completely different and more natural setting? Let's say you are a monkey who goes to a particular watering hole every day to drink. Each time you're there, you notice the other animals around, allowing you develop a sense for the various probabilities of seeing an antelope, a wildebeest, or a hippopotamus. Let the symbol a_0 denote the case where the antelope is not present and a_1 denote the case where it is. Similarly w_0, w_1, h_0, and h_1.

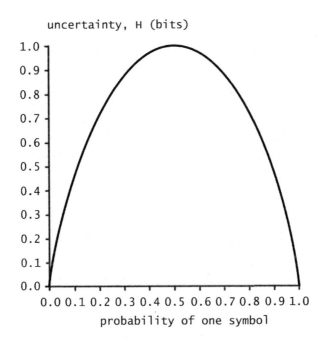

Figure A-1. Shannon entropy for a system with two symbols. Because the probabilities must sum to one, one symbol's odds determine that of the other. Entropy (a.k.a. uncertainty) is maximal when the symbols are equiprobable.

For each "symbol" (which is really an event), we have a probability p and an uncertainty $u = -\log_2 p$.

Suppose that you have observed the watering hole for 1024 days (a power of two makes the calculation simpler), and all the animals have shown up every day – the probability of any individual animal *not* appearing is no greater than 1/1024. On some random day you bring your visiting monkey-in-law with you, who has never been to this particular place. When you arrive, you notice immediately that none of the animals are there.

Translating your history into information theoretic terms, we can see that the world has transmitted to you the following string: $(a_1w_1h_1)$ $(a_1w_1h_1)$ $(a_1w_1h_1)$ $(a_1w_1h_1)$...and so on, 1024 times. On the 1025th day, the day when your in-law visits, you get the symbol $(a_0w_0h_0)$. This is very surprising to you, so your attention is fully engaged. In essence, the world has transmitted a quantity of information to you by virtue of generating that symbol. As a matter of fact, it has transmitted $-3 \cdot \log_2\left(\dfrac{1}{1024}\right) = 30 \text{ bits}$. Had all of the animals appeared, just as they usually do, you would have received only $-3 \cdot \log_2\left(\dfrac{1023}{1024}\right) = 0.0042 \text{ bits}$, a much less informative and much less surprising event. But to your visiting monkey-in-law, who has never seen this watering hole, and may therefore have a different sense of the probabilities of animals appearing, some different amount of information is transmitted. Because information varies with probabilities, and probabilities are functions of experience, information is in the eye of the beholder.

APPENDIX B

Symbolic Logic

Though this will be a technical discussion, certain details are glossed over. Acutely interested readers are referred to an outside source to for even more rigor.[49]

The first ideas to introduce are those of "propositions" and of "truth value." A proposition is simply a declarative statement, something that makes a claim. For example:

- On March 1, 2006, it rained in Seattle.
- The mass of the Earth is about $6.0 \cdot 10^{24}$ kg.
- Mary is taller than Joe.

Using a kind of algebraic shorthand, we can use variables to label these propositions. These variables have no meaning at all by themselves, just as "x" can stand for anything in mathematics. So we might say, "Let P denote the proposition that 'William Gates III is a founder of Microsoft Corp.'" Or "Let Q denote the proposition that 'Red light has a longer wavelength than blue light.'" Propositions always denote declarative statements.

The next idea is that of truth value. Any of the example propositions above could potentially be true or false. In logic-speak, we would say that, for example, proposition Q, "Red light has a longer wavelength than blue light," has a truth value of True, or just T. The proposition "2+2=5" has a truth-value of False, or just F. One of our main assumptions (and note, this assumption is itself a proposition) in this discussion will be the following:

Any proposition is either true or false, but not both, and not neither.

This assumption, called the *Law of the Excluded Middle*, conforms to most common sense understandings of declarative statements. All of the propositions above, for example, can be labeled as either true or false. Unless we really feel like nitpicking into the philosophy, we can probably agree that none are simultaneously true and false, and none are undecidable.

Of course, there are infinitely many statements that are not propositions at all. For example, "What time is it?" is not a proposition. Consequently it is not required for all statements to have a truth value under the Law of the Excluded Middle. But all acceptable *propositions* are declarative statements, and we do assume that all of them are either T or F, not both, not neither.

The next idea to introduce is the logical operator "not," usually written with the tilde symbol ~. If we have a proposition

P = "John F. Kennedy's birthday is August 2nd,"

we can denote the negation of that statement as

~P = "*It is not the case that* John F. Kennedy's birthday is August 2nd."

Of course, if P is true, then ~P is false, and vice versa. We can therefore summarize the function of the ~ operator with the following truth table:

P	~P
T	F
F	T

Our next two operators are "and" and "or." These tools allow us to construct new propositions out of smaller ones. For example, suppose we create two propositions:

P = "Mary is taller than Joe."

Q = "Joe is taller than Sam."

We can now introduce the logical "and" operator, which we denote with the symbol *, to construct a new proposition:

R = P * Q

which we can expand into:

R = P * Q = "Mary is taller than Joe *and* Joe is taller than Sam."

R itself now has a truth-value that depends on P and Q. In particular, R is true only when *both* P and Q are true. If either P or Q is false, then R is also false. We can summarize this relationship with the following truth table:

P	Q	P * Q
T	T	T
T	F	F
F	T	F
F	F	F

Now for the "or" operator. Suppose P and Q are the same as above:

P = "Mary is taller than Joe."

Q = "Joe is taller than Sam."

The "or" operator, which we denote with the symbol |, is again used to construct a new proposition:

S = P | Q

which we expand into:

S = P | Q = "Mary is taller than Joe *or* Joe is taller than Sam."

Once again, the truth value of S depends on the truth value of P and Q. Notice, however, that "or" is a little more subtle than "and." Why? Because, in English at least, the word "or" has two slightly different meanings. The first meaning is "one or the other, but not both," as in "Would you like a small or a large salary?" This is the so-called *exclusive* or, also called the "exclusive disjunction." The second meaning, the "inclusive disjunction," is understood as "either or both," as in "Would you like juice or coffee?" Logic adopts the convention that when we say "or," we *always* mean it in the latter, inclusive sense, *i.e.* "either or both."

Let's consider the consequences of that convention. In the example above, we had

S = P | Q = "Mary is taller than Joe *or* Joe is taller than Sam."

The disjunctive convention for the "or" function means that S will be true if any one of the three following conditions is the case: (1) Mary is taller than Joe; (2) Joe is taller than Sam; (3) Mary is taller than Joe *and* Joe is taller than Sam.

This relationship allows us to summarize the truth table for the "or" function as follows:

P	Q	P | Q
T	T	T
T	F	T
F	T	T
F	F	F

The reader is encouraged to pause here, if necessary, to review everything so far until it makes complete sense. It should make so much sense that it seems inevitable. Don't proceed until you understand everything so far!

We are now ready to introduce the last piece of the puzzle, which is the "implication" operator, better known as "if...then." We denote this operator with the arrow symbol → and use it in the way you probably would expect. We start with the example statement

S = "*If* it rains today, *then* my driveway will get wet."

We can immediately see that the "if...then" structure divides the statement into two pieces:

P = "It rains today"

Q = "My driveway will get wet."

With this notation, we can rewrite the original statement, using the implication operator, as "If P, then Q," or symbolically,

S = P → Q

One might expect, based on our prior experience with "and" and "or" operators, that the truth-value of S depends in some way on the truth values of P and Q. In order to analyze how, let's consider more carefully precisely what we mean by "if, then."

If I say, "If P, then Q," what I mean is "Any time P, then Q." If a circle has radius r, then it has an area of πr^2. If I'm ten feet tall, then I can dunk a basketball. If it rains, then my driveway will get wet. When P is true, I can always count on Q being true as well.

What I specifically do *not* mean is that sometimes, no matter how rarely, P is true but Q is false. If it rains but my driveway does not get wet, this particular P → Q implication is false because the causal connection is demonstrably broken. In that case we would *not* say that "If P, then Q." For P → Q to be true, there can be no exceptions to the predictive relationship.

This is such a productive and intuitive way of thinking about implication that we are actually going to define the $\rightarrow$ operator in logical terms as follows:

$$P \rightarrow Q \equiv \sim (P * \sim Q)$$

The special $\equiv$ symbol denotes *logical equivalence*. It's like a super-equality sign. To decode this statement, read it as follows: "Saying that P implies Q is the same thing as saying that it is *not* the case that I could ever find P and not Q." Of course (and this will be important below), it's quite possible that you would observe Q and not P – that's something else entirely. If I wash my car, Q is true because my driveway is wet, but this does not necessarily mean that P is true, it rained.

Once we have defined implication in this way, it is relatively straightforward to construct the entire truth table for this operator, as follows (the two rightmost columns are identical, based on the redefinition of the operator we just gave; the third and fourth columns are just intermediate values that show our work):

P	Q	~Q	(P * ~Q)	~(P * ~Q)	P → Q
T	T	F	F	T	T
T	F	T	T	F	F
F	T	F	F	T	T
F	F	T	F	T	T

Let's deconstruct each line from this table to make sure we understand its meaning.

Line 1 (omitting intermediate columns for simplicity):

P	Q	P → Q
T	T	T

This simply means that it rained (P is true), and my driveway got wet (Q is true), so we are comfortable with the correctness, *i.e.* truth, of the entire statement "If it rains, then my driveway will get wet" [(P → Q) is true.]

Line 2:

P	Q	P → Q
T	F	F

From the second example above, we decided that when P is true but Q is false (it rained but my driveway did not get wet), the implication P → Q must also be false. This is the one really unforgivable scenario. P happened but Q did not, so the whole implication is false. That's what line 2 means.

Line 3:

P	Q	P → Q
F	T	T

This surprises most students. Initially, one's instinct is to think that since the tables are turned, so to speak, since the IF part never happened, the whole implication should be thrown away. But in the car wash example, P was false and Q was true; it's just that this fact has no bearing on the claim that if it *does* rain, then (again) my driveway will be wet. So the case where P is false but Q is true does not give us any reason to deny that P → Q is true. If there is no reason to deny that P → Q is true, but we can imagine reasonable scenarios where it is true, *and* this is consistent with the definition of implication, *i.e.* that P → Q ≡ ~ (P * ~Q), then let's just call it true.

Line 4:

P	Q	P → Q
F	F	T

Given that neither P nor Q happened, what business do we have claiming to know something about the causal relationship between the two? The justification is the same as for line 3, *i.e.* that (a) there is no logical reason to reject the truth of the implication, given the irrelevant values for P and Q, and (b) that, as written, it is consistent with the definition P $\rightarrow$ Q $\equiv$ ~ (P * ~Q).

We are now in a position to return to the question from Chapter 5 that motivated this Appendix. If I am willing to unquestioningly accept some proposition, whatever it is, don't I implicitly endorse others in doing the same thing? And if their belief contradicts mine, what can we do? In other words, what is wrong with logical contradiction?

One possible position on logical contradiction is essentially tolerance. If John claims that P, and Mary claims that ~P, that is a difference of opinion, but it's a free world, let's just allow each of them to believe whatever they want. This is in fact quite a popular stance, probably made more so due to its seemingly mature and humanitarian good grace. Despite its popularity, I want to make the case that when it comes to logic, live and let live tolerance is a hideous position and one that needs to be repudiated utterly. If we tolerate a seemingly private contradiction like this, even once, our entire universe of understanding disappears instantly. Yikes! Here is what I mean.

Suppose John claims that P. Precisely what proposition P stands for doesn't matter. It could be something as benign as "Mozart had blue eyes" or something as grandiose as "The God of Abraham exists as a supernatural force at work in the world today." For our example, let's stick with the simple case: "Mozart had blue eyes." Now, let us assume that Mary disagrees with John, and claims that ~P, Mozart did not have blue eyes. In the tolerance model, both John and Mary are allowed to be right, *i.e.* both P and ~P are true.

Now we are going to play a little game. Using John's position, we know that P is true. Forget Mary for a moment and let Q stand for some other arbitrary proposition, any proposition at all. From the knowledge that P is true, we can say that the statement

  P | Q

is also true, because under the disjunctive definition of the OR operator (|), P | Q is always true if P is true:

P	Q	P | Q
T	T	T
T	F	T
F	T	T
F	F	F

Clearly, in every case where P is true (the top two rows), P | Q is also true. We know that P is true, so it is safe to say that P | Q is also true. We can be so sure of this that we don't even need to think about P at all. We *know* that P | Q is true, and we don't need to remember the details about the mechanics of how we came to that conclusion. So far, so good.

  Sticking with the truth of P | Q, suppose for a moment that P were *not* true. Even though we don't know what Q stands for, could we say anything about its truth or falseness? Well, let's see. In the cases where P is false in the table above (the bottom two rows), there are two possibilities for Q: either true or false. But in only one of those conditions (row 3) does the statement P | Q work out to be true:

P	Q	P | Q
F	T	T

We have already established that P | Q is true, so if P is false, then Q is true:

  ~P → Q

But wait a minute. We know from Mary that P is indeed false. Therefore, Q is true. The whole chain of reasoning from beginning to end, then, goes like this:

	Claim	Justification
1.	P	John is right
2.	P | Q	Definition of OR
3.	~P → Q	2, Definition of OR
4.	~P	Mary is right
5.	Q	3,4

But wait another minute! We never even specified what Q was! Q is just a free variable. So if John and Mary disagree and both are right, and we're really serious about letting them both be right, then absolutely every proposition, every claim about absolutely every subject, is true. Richard Nixon was 188 feet tall. The moon is made out of green cheese. The earth is flat. The earth is *not* flat. These are all possible values for Q and thus are all provably true statements, all due to the simple fact that John and Mary contradict each other regarding Mozart's eye color, and, because we're tolerant people, we allow both John and Mary to be right.

Although Mozart's eye color is probably checkable, this situation would have arisen no matter what the John / Mary squabble was about. Had they disagreed about the existence of an afterlife, the predictive power of astrology, or the true nature of God and we had followed the same procedure, allowing each of them to be Right, we would have unleashed an epistemological hell in which everything was true and nothing was true and everything was false and nothing was false. If there is no way to logically or empirically eliminate one of the two contradicting positions, we are guaranteed to end up in this situation.

From a logical standpoint, this is why we cannot allow *any* contradictions, anywhere. Ever.

REFERENCES

Abraham HD & Duffy FH (2001) EEG coherence in post-LSD visual hallucinations. Psychiatry Research. 107(3): 151-63.

Aghajanian GK & Sanders-Bush E (2002) Serotonin. In Davis KL, Charney D, Coyle JT, Nemeroff C (eds.) Neuropsychopharmacology: The Fifth Generation of Progress. American College of Neuropsychopharmacology. <http://www.acnp.org/Docs/G5/CH2_15-34.pdf>. Retrieved 2006-07-20.

Aghajanian GK & Marek GJ (1997) Serotonin induces excitatory postsynaptic potentials in apical dendrites of neocortical pyramidal cells. Neuropharmacology 36(4-5): 589-99.

Ammon K & Gandevia SC (1990) Transcranial magnetic stimulation can influence the selection of motor programmes. Journal of Neurology Neurosurgery and Psychiatry 53: 705-707.

Averof M & Cohen SM (1997) Evolutionary origin of insect wings from ancestral gills. Nature 385(6617): 627-631.

Bear DM & Fedio P (1977) Quantitative analysis of interictal behavior in temporal lobe epilepsy. Archives of Neurology 34(8): 454-67.

Behe M (1998) Molecular Machines: Experimental Support for the Design Inference. <http://www.discovery.org/scripts/viewDB/index.php?program=CSC&command=view&id=54>. Retrieved 2005-11-11.

Behrendt RP (2003) Hallucinations: synchronisation of thalamocortical gamma oscillations underconstrained by sensory input. Consciousness and Cognition 12(3): 413-51.

Bender L, Faretra G, Cobrinik L (1963) LSD and UML treatment of hospitalized disturbed children. Recent Advances in Biological Psychology 5: 84-92.

Benson DF (1991) The Geschwind syndrome. Advances in Neurology 55: 411-21.

Bhattacharya J, Petsche H, Pereda E (2001) Long-range synchrony in the gamma band: role in music perception. Journal of Neuroscience 21(16): 6329-37.

Bialek W, Rieke F, de Ruyter van Steveninck RR, Warland D (1991) Reading a neural code. Science 252(1014): 1854-1857.

Bliss TV, Lømo T (1973) Long-lasting potentiation of synaptic transmission in the dentate area of the anaesthetized rabbit following stimulation of the perforant path. Journal of Physiology 232(2): 331-56.

Borg J, Andrée B, Soderstrom H, Farde L (2003) The serotonin system and spiritual experiences. American Journal of Psychiatry 160(11): 1965-1969.

Brasil-Neto JP et al. (1992) Focal transcranial magnetic stimulation and response bias in a forced-choice task. Journal of Neurolology Neurosurgery and Psychiatry 55: 964–966.

Brock J, Brown CC, Boucher J, Rippon G (2002) The temporal binding deficit of autism. Development and Psychpathology 14: 209-224.

Broderick P & Phelix C (1997) Serotonin (5-HT) within dopaminergic reward circuits signals open-field behavior: II. Basis for 5-HT--DA interaction in cocaine dysfunctional behavior. Neuroscience and Biobehavioral Reviews 21(3):227-260.

Brodie M & Bunney E (1996) Serotonin potentiates dopamine inhibition of ventral tegmental area neurons in vitro. Journal of Neurophysiology 76(3):2077-2082.

Boulenguez P, Rawlins J, Chauveau J, Joseph M, Mitchell S and Gray J (1996) Modulation of dopamine release in the nucleus accumbens by 5-HT1B agonists: involvement of the hippocampo-accumbens pathway. Neuropharmacology 35(11):1521-1529.

Bourland DD, Jr. & Johnstone PD (eds.) (1991) To Be or Not: An E-Prime Anthology. (San Francisco: International Society for General Semantics).

Canolty RT, Edwards E, Dalal SS, Soltani M, Nagarajan SS, Kirsch HE, Berger MS, Barbaro NM, Knight RT (2006) High gamma power is phase-locked to theta oscillations in human neocortex. Science 313: 1626-1628.

Cook EH, Courchesne R, Lord C, Cox NJ, Yan S, Lincoln A, Haas R, Courchesne E, Leventhal BL (1997) Evidence of linkage between the serotonin transporter and autistic disorder. Molecular Psychiatry 2(3): 247-250.

Cooper JR, Bloom FE, Roth RH (1991) The Biochemical Basis of Neuropharmacology, 6th edition. (New York: Oxford University Press).

Cooper G & Meyer LB (1963) The Rhythmic Structure of Music. (Chicago: University of Chicago Press).

Copi IM (1979) Symbolic Logic. (New York: Macmillan).

Danto AC (2003) The Abuse of Beauty: Aesthetics and the concept of art. (Chicago: Carus Publishing Co.)

Dawkins R (2006) The God Delusion. (Boston: Houghton Mifflin).

de Angelis L (2002) 5-HT2A antagonists in psychiatric disorders. Current Opinion in Investigational Drugs 3(1): 106-12.

Dennett DC (2006) Breaking the Spell. (New York: Penguin Group).

Diamond J (1992) The Third Chimpanzee. (New York: HarperCollins).

Di Matteo V, Di Mascio M, Di Giovanni G, Esposito E (2000) Acute administration of amitriptyline and mianserin increases dopamine release in the rat nucleus accumbens: possible involvement of serotonin2C receptors. Psychopharmacology (Berl) 150(1): 45-51.

Eagleman DM (2004) The where and when of intention. Science 303: 1144-1146.

Enard W, Przeworski M, Fisher SE, Lai CS, Wiebe V, Kitano T, Monaco AP, Pääbo S (2002) Molecular evolution of FOXP2, a gene involved in speech and language. Nature 418(6900): 869-72.

Enquist M & Johnstone RA (1997) Generalization and the evolution of symmetry preferences. Proceedings of the Royal Society B: Biological Sciences 264(1386): 1345-48.

Fell J, Klaver P, Lehnertz K, Grunwald T, Schaller C, Elger CE, Fernández G (2001) Human memory formation is accompanied by rhinal–hippocampal coupling and decoupling. Nature Neuroscience 4(12): 1259-1264.

Fletcher PJ & Korth KM (1999) Activation of 5-HT1B receptors in the nucleus accumbens reduces amphetamine-induced enhancement of responding for conditioned reward. Psychopharmacology (Berl) 142(2): 165-74.

Fost JW (1999) Neural rhythmicity, feature binding, and serotonin: a hypothesis. The Neuroscientist 5: 79-85.

Galton F (1878) Composite portraits. Nature (May 23): 97-100.

Griffiths RR, Richards WA, McCann U, Jesse R (2006) Psilocybin can occasion mystical-type experiences having substantial and sustained personal meaning and spiritual significance. Psychopharmacology. On-line article. <http://dx.doi.org/10.1007/s00213-006-0457-5>. Retrieved 2006-07-14.

Gruber T, Müller MM, Keil A (2002) Modulation of induced gamma band responses in a perceptual learning task in the human EEG. Journal of Cognitive Neuroscience. 14: 732-744.

Haddjeri N, Blier P, de Montigny C (1998) Long-term antidepressant treatments result in a tonic activation of forebrain 5-HT1A receptors. Journal of Neuroscience 18(23): 10150-6.

Haggard P (2005) Conscious intention and motor cognition. Trends in the Cognitive Sciences 9(6): 290-295.

Hamer D (2004) The God Gene. (New York: Doubleday).

Harris S (2004) The End of Faith. (New York: Norton).

Hattox A, Li Y, Keller A (2003) Serotonin regulates rhythmic whisking. Neuron 39: 343-352.

Hebb DO (1949) The Organization of Behavior: a Neuropsychological Theory. (New York: Wiley).

Heninger GR (2000) The role of serotonin in clinical disorders. In Psychopharmacology: The Fourth Generation of Progress. American College of Neuropsychopharmacology. <http://www.acnp.org/g4/GN401000044/ CH044.html>. Retrieved 2006-07-20.

Herrmann CS, Lenz1 D, Junge S, Busch NA, Maess B (2004) Memory-matches evoke human gamma-responses. BMC Neuroscience 5: 13.

Hopfield JJ (1982) Neural networks and physical systems with emergent collective computational abilities. Proceedings of the National Academy of Sciences 79(8): 2554-2558.

Ishihara K & Sasa M (1999) Mechanism underlying the therapeutic effects of electroconvulsive therapy (ECT) on depression. Japanese Journal of Pharmacology 80(3): 185-9.

Jackendoff R (1994) Patterns in the Mind. (New York: BasicBooks).

Jacobs B (1994) Serotonin, motor activity, and depression-related disorders. American Scientist 82: 456-463.

Jacobs B & Azmitia E (1992) Structure and function of the brain serotonin system. Physiological Reviews 72:165-229.

Jacobs BL & Fornal CA (2000) Serotonin and behavior: a general hypothesis. In Psychopharmacology: The Fourth Generation of Progress. American College of Neuropsychopharmacology. <http://www.acnp.org/ g4/GN 401000044/CH044.html>. Retrieved 2006-07-20.

James W (1902) The Varieties of Religious Experience. (New York: Longmans, Green, and Co.).

Kahn D, Pace-Schott EF, Hobson JA (1997) Consciousness in waking and dreaming: the roles of neuronal oscillation and neuromodulation in determining similarities and differences. Neuroscience 78(1): 13-38.

Kaiser J, Hertrich I, Ackermann H, Mathiak K, Lutzenberger W (2005) Hearing lips: gamma-band activity during audiovisual speech perception. Cerebral Cortex 15(5): 646-653.

Keil A, Müller MM, Ray WJ, Gruber T, Elbert T (1999) Human gamma band activity and perception of a Gestalt. The Journal of Neuroscience 19(16): 7152-7161.

Kierkegaard S (1985) Fear and Trembling. Trans. A. Hannay, (London: Penguin).

Klein R & Edgar B (2002) The Dawn of Human Culture. (New York: Wiley).

Knoblauch A & Palm G (2001) Pattern separation and synchronization in spiking associative memories and visual areas. Neural Networks 14(6-7): 763-80.

Korzybski A (1933) Science and Sanity: An Introduction to Non-Aristotelian Systems and General Semantics (International Non-Aristotelian Library). (Institute of General Semantics, 5th edition, 1995).

Kosmin BA, Mayer E, Keysar A (2001) American Religious Identification Survey. <http://www.gc.cuny.edu/

faculty/research_briefs/ aris/aris_index.htm>. Retrieved 2006-06-21.

Kovacs MG (1990, trans.) The Epic of Gilgamesh. (Stanford: Stanford University Press).

Kraut MA, Calhoun V, Pitcock JA, Cusick C, Hart J Jr (2003) Neural hybrid model of semantic object memory: implications from event-related timing using fMRI. Journal of the International Neuropsychological Society 9(7): 1031-40.

Kurzweil R (1999) The Age of Spiritual Machines. (New York: Viking Adult).

Labossiere, MC (1995) Fallacies. <http://www.nizkor. org/features/ fallacies/>. Retrieved 2006-06-15.

Langlois JH & Roggman LA (1990) Attractive faces are only average. Psychological Science 1:115-121.

Lee KH & McCormick DA (1996) Abolition of spindle oscillations by serotonin and norepinephrine in the ferret lateral geniculate and perigeniculate nuclei in vitro. Neuron 17: 309–321.

Lerner L (2005) What should we think about Americans' beliefs regarding evolution? Skeptical Inquirer (November issue).

Libet B et al. (1983) Time of conscious intention to act in relation to onset of cerebral activity (readiness-potential). The unconscious initiation of a freely voluntary act. Brain 106: 623–642.

London ED, Waller SB, Vocci FJ, Buterbaugh GG (1982) Age-dependent reduction in maximum electroshock convulsive threshold associated with decreased concentrations of brain monoamines. Pharmacology Biochemistry and Behavior 16(3): 441-7.

Lutz A, Greischar LL, Rawlings NB, Ricard M, Davidson RJ. (2004) Long-term meditators self-induce high-amplitude gamma synchrony during mental practice.

Proceedings of the National Academy of Sciences. 101: 16369-73.

Lutzenberger W (2003) Induced gamma-band activity and human brain function. The Neuroscientist 9(6): 475-484.

Maclean JN, Cowley KC, Schmidt BJ (1998) NMDA receptor-mediated oscillatory activity in the neonatal rat spinal cord is serotonin dependent. Journal of Neurophysiology 79: 2804-2808.

Mitchell DR (2006) "The Evolution of Eukaryotic Cilia and Flagella as Motile and Sensory Organelles." In: Origins and Evolution of Eukaryotic Endomembranes and Cytoskeleton, edited by Gáspár Jékely.

Muramoto O (2003) The role of the medial prefrontal cortex in human religious activity. Medical Hypotheses 62(4): 479-485.

Newberg A & d'Acquili E (2000) The neuropsychology of aesthetic, spiritual, and religious states. Zygon 35(1): 39-51.

Newberg A, d'Aquili E, Rause V (2001) Why God Won't Go Away: Brain science and the biology of belief. (New York: Ballantine).

Newman J & Grace AA (1999) Binding across time: the selective gating of frontal and hippocampal systems modulating working memory and attentional states. Consciousness and Cognition 8(2): 196-212.

Nocjar C, Roth BL, Pehek EA (2002) Localization of 5-HT(2A) receptors on dopamine cells in subnuclei of the midbrain A10 cell group. Neuroscience 111(1): 163-76.

Ogata A & Miyakawa T (1998) Religious experiences in epileptic patients with a focus on ictus-related episodes. Psychiatry and Clinical Neurosciences 52: 321-325.

Pape HC & McCormick DA (1989) Noradrenaline and serotonin selectively modulate thalamic burst firing by enhancing a hyperpolarization-activated cation current. Nature 340: 715-718.

Pavlova M, Lutzenberger W, Sokolov A, Birbaumer N (2004) Dissociable cortical processing of recognizable and non-recognizable biological movement: Analysing gamma MEG activity. Cerebral Cortex 14:181-188.

Penrose R (1994) Shadows of the Mind. (New York: Oxford University Press).

Perrett DI, Lee KJ, Penton-Voak I, Rowland D, Yoshikawa S, Burt DM, Henzi SP, Castles DL, Akamatsu S (1998) Effects of sexual dimorphism on facial attractiveness. Nature 394: 884-887.

Persinger MA (1983) Religious and mystical experiences as artifacts of temporal lobe function: A general hypothesis. Perceptual and Motor Skills 57: 1255-1262.

Pew Research Center (2002) "Americans struggle with religion's role at home and abroad." Forum on Religion & Public Life Survey Results.

Pew Charitable Trusts (2004) "Religion and the Environment." Forum on Religion & Public Life Survey Results.

Pinker S (1994) The Language Instinct. (New York: HarperCollins).

Posada A, Hugues E, Franck N, Vianin P, Kilner J (2003) Augmentation of induced visual gamma activity by increased task complexity. European Journal of Neuroscience 18: 2351-2356.

Raffone A & van Leeuwen C (2003) Dynamic synchronization and chaos in an associative neural network with multiple active memories. Chaos 13(3):1090-1104.

Ramachandran VS & Blakeslee S (1998) Phantoms in the Brain. (New York: William Morrow).

Ramachandran V, Hirstein W, Armel K, Tecoma E and Iragui V (1997) The neural basis of religious experience. Society for Neuroscience Abstracts 23(2):1316.

Ramachandran V & Hirstein W (1999) The science of art: A neurological theory of aesthetic experience. Journal of Consciousness Studies 6(6): 15-51.

Rhodes G & Tremewan T (1996) Averageness, exaggeration, and facial attractiveness. Psychological Science 7(2): 105-110.

Rhodes G & Zebrowitz LA (2002) Advances in Visual Cognition, Vol. 1. Facial Attractiveness: Evolutionary, Social, and Cognitive Perspectives. (Westport, CT: Ablex).

Ribeiro-Do-Valle L, Fornal C, Litto W and Jacobs B (1989) Serotonergic dorsal raphe unit activity related to feeding/grooming behaviors in cats. Society for Neuroscience Abstracts 15:1283.

Rubenstein AJ, Kalakanis L, Langlois JH (1999) Infant preferences for attractive faces: A cognitive explanation. Developmental Psychology 35(3): 848-855.

Rubenstein AJ, Langlois JH, Roggman LA (2001) What makes a face attractive and why: The role of averageness in defining facial beauty. In G. Rhodes and L. Zebrowitz (eds.) Advances in Visual Cognition.

Rudnick G & Wall S (1992) The molecular mechanism of "ecstasy" [3,4-methylenedioxymethamphetamine (MDMA)]: serotonin transporters are targets for MDMA-induced serotonin release. Proceedings of the National Academy of Sciences 89: 1817-1821.

Sartre JP (1956) Being and Nothingness. Trans. Hazel E. Barnes. (New York: Washington Square Press).

Saver JL & Rabin J (1997) The neural substrates of religious experience. J Neuropsychiatry Clinical Neuroscience 9: 498-510.

Schneider TD (2005) Information Theory Primer. <http://www.ccrnp.ncifcrf.gov/~toms/paper/primer/primer.pdf>. Retrieved 2006-06-20.

Sagan C (1996) The Demon-Haunted World. (New York: Ballantine).

Sheard MH & Aghajanian GK (1967) Neural release of brain serotonin and body temperature. Nature 216: 495-496.

Shermer M (2000) How We Believe. (New York: Henry Holt).

Sillar KT & Simmers AJ (1994) 5HT induces NMDA receptor-mediated intrinsic oscillations in embryonic amphibian spinal neurons. Proceedings of the Royal Society of London Series B Biological Sciences 255(1343): 139-45.

Sillar KT, Wedderburn JF, Simmers AJ (1992) Modulation of swimming rhythmicity by 5-hydroxytryptamine during post-embryonic development in Xenopus laevis. Proceedings of the Royal Society of London Series B Biological Sciences 250(1328): 107-14.

Sirigu A et al. (2004) Altered awareness of voluntary action after damage to the parietal cortex. Nature Neuroscience 7: 80–84.

Solso RL (1996) Cognition and the Visual Arts. (Cambridge: MIT Press).

Stockmeier CA, Shapiro LA, Dilley GE, Kolli TN, Friedman L, Rajkowska G (1998) Increase in serotonin-1A autoreceptors in the midbrain of suicide victims with major depression-postmortem evidence for decreased serotonin activity. Journal of Neuroscience 18(18): 7394-401.

Tegmark M (2000) The importance of quantum decoherence in brain processes. Physical Review E 61: 4194-4206.

Teitler M, Leonhardt S, Appel N, De Souza E and Glennon R (1990) Receptor pharmacology of MDMA and related hallucinogens. Annals of the New York Academy of Sciences 600: 626-638.

Theunissen F, Roddey JC, Stufflebeam S, Clague H, Miller JP (1996) Information theoretic analysis of dynamical encoding by four identified primary sensory interneurons in the cricket cercal system. Journal of Neurophysiology 75: 1345-1364.

Tor-Agbidye J, Yamamoto B, Bowyer JF (2001) Seizure activity and hyperthermia potentiate the increases in dopamine and serotonin extracellular levels in the amygdala during exposure to d-amphetamine. Toxicological Sciences 60: 103-111.

Tucker DM, Novelly RA, Walker PJ (1987) Hyperreligiosity in temporal lobe epilepsy: redefining the relationship. Journal of Nervous and Mental Disease 175(3): 181-4.

US Department of Justice (2003). Press release. Justice department closes religious discrimination inquiry at Texas Tech University. <http://www.usdoj.gov/opa/pr/2003/April/03_crt_247.htm>. Retrieved 2006-06-15.

Vanhaereny M, d'Errico F, Stringer C, James SL, Todd JA, Mienis HK (2006) Middle Paleolithic shell beads in Israel and Algeria. Science 312(5781): 1785-1788.

Villalba R & Harrington C (2003) Repetitive self-injurious behavior: the emerging potential of psychotropic intervention. Psychiatric Times 20(2).

Volkmar F & Nelson D. Seizure disorders in autism. Journal of the American Academy of Child and Adolescent Psychiatry 29(1): 127-129.

Ward L (2003) Synchronous neural oscillations and cognitive processes. Trends in the Cognitive Sciences 7(12): 553-559.

Waxman S & Geschwind N (1975) The interictal behavior syndrome of temporal lobe epilepsy. Archives of General Psychiatry 32(12):1580-1586.

Welsh JP, Placantonakis DG, Warsetsky SI, Marquez RG, Bernstein L, Aicher SA (2002) The serotonin hypothesis of myoclonus from the perspective of neuronal rhythmicity. Advances in Neurology 89: 307-29.

White S, Obradovic R, Imel K and Wheaton M (1996) The effects of methlyenedioxymethamphetamine (MDMA, "ecstasy") on monoaminergic neurotransmission in the central nervous system. Progress in Neurobiology 49(5):455-479.

Whitfield TWA & Slatter PE (1979) The effects of categorization on a furniture selection task. British Journal of Psychology 70: 65-75.

Wilson EO (1998) Consilience. (New York: Alfred A. Knopf).

Wittgenstein L (1921) Tractatus Logico-philosophicus. Trans. D. F. Pears and B. F. McGuiness. (New York: Routledge).

Wittgenstein L (1958) Philosophical Investigations, 3rd edition. Trans. G. E. M. Anscombe. (Englewood Cliffs, NJ: Prentice-Hall).

Wong V (1993) Epilepsy in children with autistic spectrum disorder. Journal of Child Neurology 8(4): 316-322.

Zeki S (1999a) Inner Vision. (New York: Oxford University Press).

Zeki S (1999b) Art and the Brain. Journal of Consciousness Studies 6(6): 76-96.

IMAGE CREDITS

Figure 1-4: Reproduced with permission,
© www.faceresearch.org

Figure 2-1: Adapted from Yarbus AL (1967) Eye movements
during perception of complex objects, in L. A. Riggs,
(ed.), Eye Movements and Vision, Plenum Press, New
York, with kind permission of Springer Science and
Business Media.

Figure 2-2: Reproduced with permission from Solso RL (1996)
"Cognition and the Visual Arts," p.95, MIT Press,
Cambridge, MA.

Figure 2-3: Reproduced with permission from Solso RL (1996)
"Cognition and the Visual Arts," p.99, MIT Press,
Cambridge, MA.

Figure 2-6: © Vienna Natural History Museum.

Figure 2-7: Haggin Museum.

Figure 2-8: © Norman Rockwell.

Figure 2-9: Reproduced from "Daily Headlines," 2006-10-31. ©
University of Arkansas.

Figure 2-10: Haags Gemeentemuseum, The Hague.

Figure 3-1: Adapted from Behe (1998).

NOTES

[1] It can also pay off if the creature is smart enough to make exploration safe. This may be why curiosity and intelligence are correlated.

[2] You can also average together multiple different photos of the same person. This way, small blemishes won't be in exactly the same place and will disappear the same way they would for photos of different people, and you get the same "soft focus" effect that you do with straight photographic averaging (but not morphing). It turns out that such single-face averages are not significantly more attractive than just a regular un-averaged photo of that person.

[3] For neuroscience beginners: neurons are electrical. The way they communicate with each other is by generating sharp spikes of electricity called "action potentials" or just "spikes." Every time a spike is generated, we say the cell "fires." These spikes propagate very rapidly throughout the cell. When they reach the end of a cell, some kind of neurotransmitter is dumped into the synapse. That neurotransmitter then binds with special receptors on the next cell, causing it to either increase or decrease its rate of generating spikes. Looking at the activity of a cell over time yields a plot called the "spike train."

[4] Proposals include the activation of reciprocal connections between synchronized areas, or involvement of a third brain structure (like the thalamus) as a central coordinator.

[5] As measured by the root-mean-square (RMS) value of the autocorrelogram.

[6] Solso (1996), pp. 242-243.

[7] Sheard & Aghajanian (1967).

[8] The neurophysiology of serotonin is extremely complex. There are around 14 separate receptors for serotonin, some of them presynaptic and some postsynaptic, some both. The effects of each receptor can differ depending on where it is in the brain. Classical hallucinogens, including LSD, activate 5-HT_2 receptors (of which there are three subtypes), while MDMA actually causes serotonin and dopamine release. MDMA also has effects on dopamine receptors and on the interactions between serotonin and dopamine.

It is MDMA's direct dopaminergic effects that probably account for the euphorigenic quality of the drug. Meanwhile, there are known connections between other serotonin receptors and abnormal rhythmic neural activity: mice lacking the 5-HT1c receptor experience spontaneous seizures, and in humans, 5-HT1a binding is decreased in patients with temporal lobe epilepsy.

[9] This is controversial, with suggestive evidence on both sides. On one hand, the anthropological evidence for a discontinuity in cultural sophistication is good. Also, genetic findings such as Enard *et al.* (2002) support the plausibility of discrete mutations playing large roles in linguistic ability. On the other hand, there are some findings of symbolic capacity significantly before the 50 ka point – see Vanhaereny *et al.* (2006) for an example. Finally, Stephen Pinker notes that even today there is no such thing as a generalized tendency toward symbolic processing that includes language, art, and ritual. A modern three-year old child, he hints, performs well linguistically but is not especially capable of artistic or ritualistic symbolism.

[10] *The Language Instinct*, p. 153.

[11] *The Language Instinct*, pp. 367-368.

[12] Quoted in Zeki (1999), p. 170.

[13] Ramachandran & Hirstein (1999), pp. 31-32.

[14] This is precisely the part of the process that is broken in Capgras syndrome. One might predict, therefore, that Capgras patients would be immune to the prototypicality effect, *i.e.* they would not find prototypical faces more attractive.

[15] Zeki, *Inner Vision*, p. 11-12.

[16] Paraphrased from Gregory Bateson in *The Age of Spiritual Machines*, by Ray Kurzweil, p.159.

[17] "Entities should not be multiplied beyond necessity."

[18] *Phantoms in the Brain*, pp. 180-181.

[19] For a very well executed recent example, see Griffiths *et al.* (2006).

[20] This is what Newberg and his colleagues call the Absolute Unitary Being. They also believe that this state is brought about by rhythmic stimulation, but in their opinion, the ultimate origin is not in sensory binding but rather in the neural circuitry of sexual experience (*Why God Won't Go Away*, pp. 120, 125).

[21] In an etymological twist, the word "symbol" derives from the Greek *synballein*, from *syn-* "together" and *ballein* "to throw," and originally held a connotation of favorable comparison. The word "devil" derives from the almost antonymic Greek *diaballein*, from *dia-* "across, through" and again *ballein* "to throw." Contra "symbol," this came to connote slander, as in Latin's *diabolus*, and then evil spirit, as in Old English's *diafol*. We also find the Word of God, the Prince of Lies, and name as power. This timeless triad of humanity, language, and the supernatural is clearly a conceptual strange attractor.

[22] *Why God Won't Go Away*, pp. 87-125.

[23] *Breaking the Spell*, p. 259.

[24] See for example Tegmark (2000).

[25] Jupiter Scientific (2004) An estimate of the number of Shakespeare's atoms in a living human being. <http://jupiterscientific.org/review/shnecal.html>. Retrieved 2006-06-16.

[26] Whether this auto-self-discovery is inevitable is a matter for debate. Evolutionists like Stephen Jay Gould emphasize the historical contingency and improbability of any particular biological outcome. According to that view, there was no "inexorable march" toward the uniquely symbolic and theoretical nature of the human mind. It was only a freak occurrence, after all, that displaced the dinosaurs and cleared the way for the age of mammals. On the other hand, causalist physicalism would seem to indicate that although there was no teleological guidance toward things as they stand today, it was in some sense destined to happen anyway because of the determinist nature of all the component physics. On the third hand (first foot?) the Copenhagen interpretation of quantum mechanics would seem to indicate that the universe is not causal, in which case Stephen Jay Gould was right – it could equally have gone lots of other ways. Whether some other symbolic intelligence would emerge, somewhere in the universe, and formulate the same natural theories as we have is another question. Finally, the Copenhagen interpretation is debated by physicists (not just philosophers) and is potentially replaceable by Bohmian mechanics – a version of quantum mechanics that rescues determinism. From a theological perspective, some of Teilhard's views of the Omega Point might come to mind...though this mention of him should not be interpreted as an endorsement.

[27] See Averof & Cohen (1997) as a starting point.

[28] For an extensive discussion of the evolution of the cilium and its functional (but non-motile) precursors, see Mitchell (2006).

[29] *Being and Nothingness*, p. 101.

[30] Quote is from Wikipedia <www.wikipedia.org>, "Being and Nothingness," retrieved 2006-06-16.

[31] I am not trying to explain Sartre's philosophy in its entirety, nor am I particularly claiming that his system is good; for example I utterly refute his premise that we are free. However, it *feels* like we are free, and it is in that world of appearances that *mauvaise foi* has explanatory power. Moreover, the mere status of his as an atheist philosophy makes it relevant here.

[32] Zeki (1999b), pp. 81-82.

[33] See also Nietzsche's idea that it is only by overcoming ourselves on the road to the übermensch do we achieve anything of value.

[34] See for example *Phaedo*, Stephanus pages 76a-80d and 110b-114c.

[35] We find an intriguing connection here to Alfred Korzybski's theory of General Semantics, in which he identified specific forms of the verb "to be" as semantically troubling, and to the subsequent work of D. David Bourland Jr. Bourland instantiated Korzybski's observation by inventing E-Prime, the self-conscious dialect of English that eliminates all forms of the verb "to be." Following Korzybski, Bourland did not view the existential use of "to be," as in "There is an apple on the table" as especially troubling. He focused instead on identity ("That is an apple") and predication ("The apple is red") but he seems to channel Wittgenstein in identifying the first major advantage of E-Prime as the elimination of what he calls Vanishing Questions. He said, "One simply cannot ask a number of questions – some would say pseudo-questions – that have preoccupied many people. What is man? What is woman? Is it art? What is my destiny? Who am I? Such questions, by virtue of their semantic structure, set the stage for identifications and confusions in orders of abstraction. They tend to lead to discourse in which the likelihood of useful information generation or exchange declines precipitously. One might better ask questions on a lower order of abstraction such as these: What characterizes man or woman uniquely? In what way can I relate to this art form, if any? What can

I do now to improve my future possibilities? May I have another drink?" (Bourland, 1989).

[36] It is also quite interesting that Kierkegaard was centrally concerned with whether faith is something that can be spoken about, and if it cannot be spoken about, whether this is because faith is a purely aesthetic experience and therefore something beautiful, beyond the scope of language's applicability. He says that language, as a collection of ideal forms, cannot really represent reality, and therefore cannot really be used to talk about faith. Faith is real, language is ideal, and never the twain shall meet.

[37] This is oddly and coincidentally reminiscent of the fantastical mage or the spell-casting witch, conjuring powerful and dangerous things via spoken incantations.

[38] Pew Research Center (2002), p. 2.

[39] *The Language of God.*

[40] The Harris Poll #90, December 14, 2005.

[41] *The Atlantic Online,* interview 2007-07-12 by Jennie Rothenberg Gritz. <http://www.

theatlantic.com/doc/200707u/christopher-hitchens>.   Retrieved 2007-08-20.

[42] *Breaking the Spell,* p.61.

[43] These examples are from Labossiere (1995), the best compilation of logical fallacies I know.

[44] The Civil Rights Act of 1964 includes protections for employees against religious discrimination, but in many cases students are not employees of their educational institution. In 2003, the US Department of Justice investigated a case at Texas Tech University where a professor required, in order to receive an "A" grade in the course, that students "truthfully and forthrightly affirm a scientific answer" to the question "How do you think the human species originated?" When the professor lifted the requirement in response to student complaints, the DOJ closed the investigation and stated explicitly that state-run universities had no business telling students what they should or should not believe. The DOJ also stated that religious beliefs do not need to be abandoned for scientific or medical competence.

[45] Lerner (2005).

[46] *Breaking the Spell*, p.61.

[47] The great 19th century British orator Benjamin Disraeli had a reputation as an unrelenting wit and wordsmith, and once boasted that he could make a pun on any subject. When his archrival William Gladstone dared him to make a pun on Queen Victoria, Disraeli replied, "Ah – but the Queen is not a subject."

[48] Philosophers of language were probably among the few who didn't laugh derisively when President Clinton tried to make a related point during the Monica Lewinsky investigation.

[49] See for example Copi (1979).

INDEX

Printed in Great Britain
by Amazon